SCIENCE, TRUTH, AND DEMOCRACY

OXFORD STUDIES IN PHILOSOPHY OF SCIENCE

General Editor
Paul Humphreys, University of Virginia

The Book of Evidence
Peter Achinstein

Science, Truth, and Democracy
Philip Kitcher

The Devil in the Details: Asymptotic Reasoning in Explanation, Reduction, and Emergence
Robert Batterman

SCIENCE, TRUTH, AND DEMOCRACY

Philip Kitcher

2001

OXFORD
UNIVERSITY PRESS

Oxford New York
Athens Auckland Bangkok Bogotá Buenos Aires Cape Town
Chennai Dar es Salaam Delhi Florence Hong Kong Istanbul Karachi
Kolkata Kuala Lumpur Madrid Melbourne Mexico City Mumbai Nairobi
Paris São Paulo Shanghai Singapore Taipei Tokyo Toronto Warsaw

and associated companies in
Berlin Ibadan

Published by Oxford University Press, Inc.
198 Madison Avenue, New York, New York 10016

Library of Congress Cataloging-in-Publication Data
Kitcher, Philip, 1947–
Science, truth, and democracy / Philip Kitcher.
p. cm.—(Oxford studies in philosophy of science)
ISBN 0-19-514583-6
1. Science—Social aspects. 2. Science—Philosophy. I. Title. II. Series.
Q175.5 .K525 2002
501—dc21 2001036144

3 5 7 9 8 6 4 2

Printed in the United States of America
on acid-free paper

In memory of Roger
who taught so many so much about friendship

Acknowledgments

THE IDEAS PRESENTED IN WHAT FOLLOWS have emerged from a gradual rethinking of the general account of the sciences I offered in earlier discussions (particularly in *The Advancement of Science*) in light of my studies of the uses of scientific findings in social contexts. In particular, reflection on the character and likely employment of current and foreseeable work in molecular genetics has led me to appreciate points to which I was previously insensitive. Yet although this book is the result of my own intellectual meandering, I have tried to present my current views in a way that would not presuppose prior acquaintance with the literature in philosophy of science. I hope it will be accessible to people in a wide variety of disciplines and also to those who are beginning study of philosophical questions about the sciences.

I owe many debts to many people. Reviewers of *The Advancement of Science* made important points, and I am particularly indebted to Richard Boyd, John Dupré, Ian Hacking, Jarrett Leplin, Isaac Levi, Carl Matheson, and Richard Miller. An early draft of the book benefited enormously from the constructive criticism of Nancy Cartwright and William Loomis; discussions with Nancy and with Bill were most helpful in showing me how I should reshape an unnecessarily complex and forbidding manuscript. I received useful advice also from Clark Glymour, Alvin Goldman, Peter Godfrey-Smith, and Michael Rothschild. Noretta Koertge led a discussion group on that manuscript at Indiana University and sent me valuable reports on responses to the various chapters; I am grateful to her and to the others who contributed their thoughts.

During 1999, I had the opportunity to discuss this material in two seminars, one at UCSD and one at Columbia University. I would like to thank the students who helped me refine my ideas: particularly P. D. Magnus and Carl Sachs

(UCSD), and Eleonora Cresto, Ernesto Garcia, Stephanie Ruphy, and Jeremy Simon (Columbia). Central themes of the book also trace back to a seminar I taught jointly with Martin Rudwick at UCSD. Besides Martin's valuable suggestions, a number of the participants in that seminar made important points; I am particularly grateful to Kyle Stanford and Thomas Sturm.

Along the way I have been aided by numerous colleagues and members of audiences to which I've presented pertinent material. I would like to thank Richard Arneson, Jerrold Aronson, David Brink, Ruth Chang, Susan Dwyer, Delia Graff, Richard Grandy, Amy Guttman, Gilbert Harman, Richard Nelson, Helen Nissenbaum, Sherrilynn Rousch, and Peter Singer. Two colleagues deserve special mention: both Isaac Levi and Joseph Raz have helped me enormously to see how to articulate my views.

The final version has benefited greatly from the constructive suggestions of John Dupré, Paul Humphreys, and an anonymous reader for Oxford University Press. It was also considerably improved by Patricia Kitcher's wise counsel.

I apologize in advance if I have forgotten anyone who corrected my mistakes or pointed me in a useful new direction. And it should go without saying that none of the kind people who have helped me is responsible for the residuum of error.

Contents

Introduction

FROM TIME TO TIME, WHEN I EXPLAIN to a new acquaintance that I'm a philosopher of science, my interlocutor will nod agreeably and remark that that surely means I'm interested in the ethical status of various kinds of scientific research, the impact that science has had on our values, or the role that the sciences play in contemporary democracies. Although this common response hardly corresponds to what professional philosophers of science have done for the past decades, or even centuries, it is perfectly comprehensible. For there are large questions of the kinds just indicated, questions that deserve to be posed and answered, and an intelligent person might well think that philosophers of science are the people who do the posing and the answering. The chapters that follow are a first attempt to do just that.

Issues about the value of science, or about science and values, assume a special interest at the beginning of the twenty-first century because of the recent eruption of the "Science Wars," a controversy that typically pits two inadequate views of science against one another. With luck, this overheated debate will join the ranks of past academic disputes on which we look back with bemusement at all the fuss. But the controversy does indicate a more enduring need for a compelling perspective on the sciences and their place in democratic society—the need to answer those who thought they could guess what philosophers of science do.

I begin from two images that have dominated much public discussion. Enthusiasts for the sciences write books and articles that proclaim the search for objective knowledge as one of the crowning achievements of our species. Detractors deny the objectivity of the sciences, question our ability to attain truth and knowledge, and conclude that the sciences are instruments of oppression.

Despite the flourishing of charges and countercharges about truth and knowledge, the underlying differences turn, I believe, on assumptions about values that are never brought into the light. The first part of this book prepares the way for considering these assumptions by moving over relatively familiar philosophical terrain. I am concerned to articulate a picture of the aims and accomplishments of the sciences so that the moral and social questions can be brought into clearer focus.

In chapter 2, I try to rehabilitate the much-abused notion of truth within the framework of a modest realism. Properly understood, the suggestion that areas of science sometimes tell us the truth about nature involves neither weighty metaphysics nor deep confusion. Chapter 3 responds to the suggestion that the evidence for scientific claims is always insufficient, so that our decisions about what to believe are always crucially affected by moral, social, political, and religious values. Some readers are likely to find my claims in these chapters obvious and will be impatient with my attention to criticisms of them; such readers are encouraged to skip or skim.

The next step is to consider how we should think about the aims of the sciences. Chapter 4 calls into question the view that there is some privileged way of conceptualizing nature. In chapter 5, I offer an analogy to show how my picture of science as providing objective knowledge does not entail that there is some unique, context-independent goal toward which inquiry aims. That analogy will also suggest a quite different way of thinking about the goals of the sciences and about scientific progress. I develop that perspective in chapter 6, arguing that our scientific goals are shaped by past projects and accomplishments, and that they evolve in light of a wide range of theoretical and practical interests.

Part I (chapters 1–6) thus develops the basis on which part II faces directly the involvement of moral, social, and political values in the sciences. As we shall see in chapter 7, the view resulting from part I reveals how difficult it is to draw a distinction between science and technology. It thus undermines the idea that technology is the appropriate focus of moral appraisal, leaving science as pure. In chapter 8, I develop further the critique of science as morally and politically neutral through an extensive discussion of arguments about free inquiry as they apply to research on race and sex differences.

The next step is to work toward a more general account of the way in which properly functioning science should conform to broader values. I begin, in chapter 9, by considering the variety of standards to which one might hold inquiry, arguing that traditional philosophies of science leave out a great deal. That sets the stage, in chapter 10, for my account of the role that the sciences should play in a democratic society. I present an ideal of "well-ordered science," intended to capture what inquiry is to aim at if it is to serve the collective good. That ideal is, I admit, one representative of a family, and it would be folly for me to pretend that my account is correct. I offer it because I believe that we need some ideal of this general kind, and that it is useful to develop one in a little detail. The discussion in chapter 11 attempts to relate the very abstract framework of chapter 10 to actual discussions of the role of the sciences in society, and I try to highlight insights and oversights of some influential and thoughtful statements of science policy.

Chapters 12 and 13 extend my argument, by confronting directly the suggestion that the value of human knowledge overrides all other concerns. In scrutinizing both the claim that science has enlightened us and that that is a wonderful thing, and the counterclaim that the "progress" of science is antithetical to human well-being, I argue that both views are unwarranted. My discussion is intended to elaborate and defend the integration of the value of knowledge with moral and political values, offered in the picture of scientific inquiry I propose.

Although it is illustrated with examples, the line of argument developed in chapters 1 through 13 is abstract and general. My aim is to understand the philosophical basis for claims that are often made in discussions and disputes about science: in an earlier age, my enterprise might have been characterized as "laying the foundations" for attitudes and policies about scientific inquiry. Applying the ideals I delineate requires detailed knowledge of a large number of things about which I must confess ignorance. Nonetheless, abstract philosophizing obviously runs the risk of irrelevance, and it seems important to offer some indications of how my proposals might have implications for our far-from-ideal world. Hence I close with a chapter in which I discuss the responsibilities of scientists in the case of the example I know best, that of contemporary genomic research.

Throughout, I have tried to write both for my fellow philosophers and also for people outside philosophy who reflect on the role of the sciences in democratic societies. Accordingly I've endeavored to keep footnotes and references to specialized literatures to a minimum. As philosophers (and, to a lesser extent, historians, social scientists, and natural scientists) read these chapters, they will surely sense links to professional debates and to particular works. Including all the lines of filiation within the text or notes would produce clutter, defeating my aim to be intelligible to a broad audience. Instead I've appended an "Essay on Sources," in which I acknowledge intellectual debts, note connections, and briefly indicate differences with other authors. I hope this will prove an appropriate substitute for the bristling thicket of notes that once figured in earlier drafts.

What follows is an essay in the etymologically original sense. It should be read as an attempt to venture into areas that philosophers of science have neglected. Like most such forays it is quite likely that the way will sometimes be lost and that landmarks will be wrongly described. But if others are moved to correct my mistakes, the essay will have achieved its principal purpose.

I

THE SEARCH FOR TRUTH

ONE

Unacceptable Images

WHAT IS THE ROLE OF THE SCIENCES in a democratic society? Some people, let us call them the "scientific faithful," say this: "The sciences represent the apogee of human achievement. Since the seventeenth century, they have disclosed important truths about the natural world, and those truths have replaced old prejudices and superstitions. They have enlightened us, creating conditions under which people can lead more satisfying lives, becoming more fully rational and more fully human. The proper role of the sciences today is to continue this process, by engaging in free inquiry and by resisting attempts to hobble investigations for the sake of any moral, political, or religious agenda."

The faithful do not believe that scientific research is completely free of moral constraints. They would agree that investigators must be honest in the presentation of their findings, and they would concede that some methods of inquiry cannot be tolerated. Mindful of the appalling activities undertaken by the Nazi doctors and of the Tuskegee syphilis study (in which black men were left untreated "for the sake of science"), they recognize that the conduct of experiments cannot override human rights—or, perhaps, even the rights of some animals. However great the intellectual benefits of disentangling the roles of nature and nurture in human development, it would be morally monstrous to breed "pure lines" of children and rear them in carefully calibrated environments. So when it is claimed that inquiry must be free, what seems to be intended is that moral, political, and religious judgments should not enter into two important contexts of decision: the formulation of projects for scientific inquiry and the appraisal of evidence for conclusions. The questions investigators address should not be limited by the ideals and the fears that happen to be prevalent in human societies. Nor should we deceive ourselves by believing what we find comfortable when that belief would be undermined by available evidence. *Sapere aude* remains our proper motto.

Others think differently. On their account, the vision just sketched is a myth. Would-be debunkers believe that it is very much in the interests of those who are currently in power in affluent societies to cultivate the idea of a pure science that stands free of moral, political, and religious values, and that the myth serves as a tactic for excluding viewpoints that the powerful would like to marginalize. The minimal criticism is that decisions about which inquiries are to be pursued are always made by invoking judgments of value. Many would add to this the suggestion that there is no objective notion of "the evidence," that decisions about which "scientific conclusions" to accept are always made on the basis of moral, political, and religious values. A more extreme critique would argue that the idea that the sciences deliver to us truths about nature is another part of the myth. In the end, institutionalized science comes to seem like an effective propaganda machine, serving the interests of the elite classes and imposing its doctrines, ideals, and products on the marginalized masses in much the way that politico-religious institutions of the past managed so successfully. Science (with a capital "S") is the heir of the Catholic Church and the Party.

Neither of these images is acceptable. Each contains elements that can be used in crafting a more adequate vision. My aim is to articulate that vision. In rejecting both the image of the scientific community as secular priesthood and its polar opposite, I offer a conception of the scientist as artisan, as a worker capable of offering to the broader community something of genuine value, whose contributions can be, and should be, responsive to a much wider range of concerns than are usually taken to be appropriate. That, of course, is only a sketch. The plausibility of the full picture will depend on the details.

Let us begin again, more concretely. There are several places at which contemporary scientific research inspires reflective people to ponder the value of lines of inquiry that are proposed and to invoke one of the images I have characterized as unacceptable. Without any suggestion that all the nuances of complex debates will be captured, it will be worth reviewing a few examples.

Consider first the Human Genome Project. Governments throughout the affluent world, but most particularly in the United States, have contributed large sums of money so that researchers will map and sequence the genome of our species (or, more exactly, a bundle of segments of DNA drawn from a small number of human beings) and the genomes of certain carefully selected other organisms. Public defense of the project often emphasizes the medical benefits that will flow from the expansion of detailed knowledge about human genes. Privately, policy-makers and politicians who favor the project talk more frequently of the economic benefits of engaging in it, the advantages of building or maintaining a lead in biotechnology, while scientific researchers, away from the microphones and cameras, point out the ways in which a huge archive of sequence data will help the "basic biology" of the new century. All this is readily comprehensible. For the economic consequences and the consequences for biological research are far more definite than the nebulous payoffs for human health.

It is already clear from cases in which we have achieved molecular insights about the causes of disease that there may be no obvious way to apply those insights in treating, curing, preventing, or ameliorating the malady in question. The molecular details underlying sickle-cell anemia have been known for half a century without yielding any successful strategies for tackling this disease. Yet there are stories of small advances. Thanks to our ability to identify alleles implicated in cystic fibrosis, it is now possible to diagnose children more quickly and to use techniques that reduce the frequency and intensity of the crises to which those who have the disease are subject. So long as one emphasizes improvements in diagnostic testing and partial gains in coping with some diseases, it is quite reasonable to claim that the genomes project (as it would more aptly be called) can bring some medical benefits. Furthermore, as we look into the future, enhanced understanding of basic biology *may* bring, several decades or a century hence, significant breakthroughs in the treatment or prevention of diseases that cause suffering and premature death for millions. It would be unwise either to rule out that possibility or to stake the (research) farm on it.

Unfortunately, as should by now be abundantly obvious, the explosion of genetic knowledge will have immediate consequences of a much darker kind. Within a decade, biotechnology companies will be offering hundreds, if not thousands, of predictive genetic tests. Given the character of the practice of medicine in much of the affluent world, it is highly likely that a significant number of people will confront information that is psychologically devastating, or be excluded from a job on genetic grounds, or be denied insurance through genetic discrimination, or face an acute dilemma about continuing a pregnancy. These consequences have been amply discussed by knowledgeable and well-meaning people, and, in all cases except the last, the solutions, in principle, to the problems are not too hard to find. Nevertheless, over a decade after the genomes project began, virtually nothing has been done to alleviate readily foreseeable harms. That fact is especially noteworthy, given the decision, made at the beginning of the project, to undertake a thorough exploration of its ethical, legal, and social consequences. In the United States that commitment was expressed in setting aside a small percentage of the (very large) funds expended on the project, and many of the suggestions for avoiding the difficulties of the new age of genetic testing come from research that has been supported in this way.

Turn to a second example. At about the same time that molecular biologists were persuading the U.S. Congress to fund the genomes project, an extremely prestigious group of physicists failed in their attempt to obtain public money to build the superconducting supercollider. In this case, the character of the public defense coincided with the private justifications given by the scientists involved. The request for a sum an order of magnitude larger than that expended on mapping and sequencing genomes was to build a facility in which minute constituents of matter could be smashed into one another at velocities considerably greater than those ever previously achieved, in the hope of discovering a rare and evanescent product of the collisions. Some politicians were probably swayed by the thought that their local constituents would benefit from jobs created by the project, but, for the majority, the decision turned on whether a considerable sum of public money should be spent in the hope of confirming and developing an esoteric theory about the ultimate constituents of matter. Physicists were eloquent in explaining how the facility they proposed was needed to continue probing the character of fundamental particles, how their planned investigations extended a line of inquiry that had given rise, successively, to the atomic theory, to conceptions of atoms as composed of elementary particles (electrons, protons, neutrons), to the discovery of quarks and the partial unification of accounts of the basic physical forces, but those who held the purse-strings were eventually unmoved by dreams of a final theory, perhaps viewing the accelerator as an expensive plaything that would generate nothing outsiders could appreciate or understand. They judged other demands to be more urgent.

In other instances, a line of proposed scientific research may be evaluated not as insufficiently beneficial but as genuinely harmful. For at least a century, the general public has been periodically informed that careful biological investigations have revealed unpleasant truths about the natural differences among members of particular groups. Inequalities in performance with respect to tasks that are socially valued have been unmasked as the result of unmodifiable characteristics, and, more or less regretfully, the investigators and those who have popularized their findings have maintained that any policy of eradicating inequalities is doomed to failure. No matter how hard we may try, there are limits to our power to boost I.Q. or to make the upper echelons of the professions available to groups that have been historically disadvantaged (people with two X chromosomes or with a tendency to produce melanin in their skin cells). In many instances, those who champion this kind of research claim that the problems they are addressing are too important to ignore, and that an enlightened social policy must be based on an awareness of the fixed obstacles that block paths we would like to take. When these defenses are challenged, the advocates can fall back on the importance of knowledge in general, and of self-knowledge in particular, independent of any practical consequences. Their opponents sometimes argue that research of so sensitive a kind must be held to stringent standards of evidence, and that socially consequential claims should not be accepted on what they see as the flimsy reasoning being offered. More fundamentally, they may draw from the dismal history of efforts to trace a biological basis for social inequality the conclusion that we have good reason to believe the appropriate standards of evidence to be simply unattainable. Taking an even more radical step, they may suggest that, even if true, these are not matters about which we should want to gain knowledge. The proposals for more research on differences due to sex, race, or class thus face the charge that the envisaged inquiries are morally suspect.

The long sequence of investigations to which I have just alluded gives rise to moral debate because the acceptance of some scientific doctrines would affect the lives of people in very obvious ways. My final example steps away from the mundane consequences, the everyday shocks that types of human beings are differentially heir to. A commonplace about the growth of the sciences is that, at various times, a new proposal has profoundly disturbed reflective people, causing them to re-evaluate, and even abandon, some of the central beliefs that have given shape and significance to their lives. The impact of Darwin's ideas on human aspirations and self-conceptions is reflected in his first disclosure of his theory to his close friend Joseph Dalton Hooker: "It is like confessing a murder," he wrote (and he meant it). Even today, of course, people continue to resist the claim that there is overwhelming evidence that Darwin was right about the history of life, and their struggles with his doctrines often take the form of conjuring a conspiracy against religion and suggesting that this is a place in which science has been distorted by prejudice. Ironically, the conception of Darwinism itself as a religion masquerading as science is not far from some academic suggestions that *all science* is permeated by prejudices and social values, concretely expressed in the example of evolutionary theory by claiming that we should understand Darwin's triumph not in terms of his evidence and sound arguments but in his ability to resonate the values of competitive, Victorian, bourgeois capitalism.

The scientific faithful have familiar ways of responding to the issues posed by the examples I have offered. Consider, first, the genomes project. The beginning of wisdom, the faithful will insist, is to distinguish sharply between science and technology. There are scientific findings about the relative positions of genetic loci on chromosomes and about the structures of the alleles at those loci, and there are technological applications of those findings within agriculture, medicine, criminology, and other social ventures. Science proposes, society disposes. On one forthright conception of the proper role of the sciences, although we may appraise the moral status of technological ventures, the scientific research itself is entirely neutral. So resolute a stance might provoke doubts once it is recognized that there are extreme cases in which it appears that *any* application of a piece of research, within the kinds of societies we can plausibly envisage, would prove destructive: consider, for example, research that reveals how cooking just the right combination of broccoli, bananas, and bluefish (or other readily obtainable ingredients) would generate an explosion that would make Hiroshima and Nagasaki appear as damp squibs. When the only consequences of applying a scientific result are so clear, and so clearly awful, then even the faithful may allow that there is a moral imperative to desist. Yet this, they are likely to suggest, is truly an extreme case. For virtually all scientific research, the consequences are unpredictable and the harms and benefits of technological applications incalculable. In such circumstances the value of the knowledge, for its own sake and for the sake of future developments to which it may lead, should prove decisive. So, in the particular instance of the genomes project, we cannot say in advance what the balance of good and bad results will be. Scientists act responsibly in gaining deeper biological knowledge and in deferring to others the problem of making the best use of what they find.

An extension of the same line of argument portrays the decision not to fund the superconducting supercollider as myopic. The value of a scientific inquiry cannot be identified just by considering the set of technological applications to which it gives rise. To discover the Higgs vector boson (the elusive particle that the apparatus was designed to hunt) would be to take a further step in the great intellectual adventure of uncovering the structure of matter. Independently of any practical spinoffs from the experiments, achieving a clearer picture of the fundamental constituents of the universe would be worthwhile for its own sake, just as it is valuable for us to know the major characteristics of our galaxy, the processes that formed our planet, and the history of life on earth. Not only do such cognitive accomplishments vastly outweigh the kinds of pragmatic concerns that figure in budgetary decisions, but they often point in unexpected ways toward future lines of scientific research that will ultimately bequeath to our descendants a vastly wider range of practical options. As the faithful like to emphasize, the history of science is full of examples in which work that initially appeared to lack any practical value proved to be crucial for subsequent developments that spawned a host of welcome technological applications: abstract approaches to computation gave birth to the word processor and the internet; breeding experiments on fruit flies eventually yielded medical genetics; and so it goes. The decision against the supercollider both ignored the intrinsic benefit of the knowledge it would bring and forgot the historical lesson that the pursuit of fundamental science brings long-term dividends.

Elaborating the argument still further, the faithful approach my third and fourth examples. They recognize that, in the short term, the articulation of unpleasant scientific truths may cause pain and suffering, and may even affect most those who have been victims of discrimination in the past. It is important, they will agree, to do whatever can be done to ensure that findings about human nature are translated into social policy in ways that are sensitive to the needs of the disadvantaged. Yet to produce an enlightened social policy we require the clearest possible account of human beings and their needs. As in other scientific cases, there is an intrinsic benefit from arriving at knowledge—perhaps an especially rich benefit when the subject is ourselves—but here there is also the practical gain of an ability to design social institutions that does not go astray because of illusory hopes. Much as we might like to believe particular things, we gain from knowing the truth about ourselves and from putting our knowledge to work.

My final example encourages a development of the theme. Even when there is no practical benefit from applying our new self-knowledge in social policies and even when the knowledge may deprive us of comforting illusions, we are still better for having it. Human beings participate in a common enterprise of fathoming nature, and that enterprise is one of the chief glories of our species. Or, to put the point differently, to shun knowledge because it might appall us is to betray an important aspect of our humanity.

In the responses I have put in the mouths of the scientific faithful, we find a number of philosophical theses. The sciences can provide us with knowledge of nature. They have a definite aim, namely to offer knowledge that is as systematic and complete as possible. That knowledge can be used for practical ends, but the moral appraisal of the uses is properly directed at technology and public policy, not at science itself. Besides its practical benefits or harms, the knowledge has intrinsic value, and that value typically overrides mundane practical concerns.

The most popular recent criticisms of "scientism" focus on the first thesis. In many academic circles, it has become increasingly popular to deny the claims of the sciences to yield knowledge (truth, we are often told, is either unattainable or a notion that is passé). As I shall try to show, this is an unfortunate way to join the debate, for the serious concerns about the credo of the faithful should focus on the subsequent theses. Can we really make sense of the idea that sciences have a single definite aim? Can we draw a morally relevant distinction between science and technology? Can we view the kind of knowledge achieved by the sciences as having overriding value?

My eventual aim is to address these questions. First, however, we must become clearer about the notions of truth, knowledge, and objectivity. So I begin with them.

TWO

The World as We Find It

MANY REFLECTIVE PEOPLE READ with a sense of bemusement the obituaries so frequently (and so gleefully) written for scientific realism. Modesty comes naturally. We believe that we cannot simply make up the world to our own specifications. Yet optimism and a sense of ambition are almost as automatic. Complicated though nature may be, it does not seem impossible that we should fathom its recondite secrets, and, as we contemplate past inquiries, we seem to discern a progressive history of past successes. The result is a simple form of scientific realism: Scientific inquiry discloses truths about a world that is independent of human cognition, and, among those truths, some do not merely identify superficial aspects of nature but reveal things and processes that are remote from everyday observation.

Simple scientific realism invites jeers. Inquiry cannot deliver truth—at least, not real Truth. At best, we can know the observable surfaces of nature. The notions of "the world" or of its "independence" from human thought are deeply confused. Worlds are as we make, or "construct," them. In many fashionable discussions, a pejorative name is enough, "Realist!" or (worse) "Positivist!"

Why, then, does scientific realism feel so close to plain sense? The reasons are not hard to find. Everyday consideration of the statements people hear and read involves appraising those statements as true or false, and, without any weighty metaphysics, it is easy to align the standard of truth with an account as old as the Greeks. The statements we care about typically contain terms that pick out constituents of the world and attribute properties and relations to them; those statements are true when the entities so identified have the properties and relations ascribed to them, and false when they do not. Although the sciences frequently discuss things and processes remote from ordinary observation—too small to see with the naked eye, or too distant—and despite the fact that they often advance theses of great generality, it is not obvious why their intricate methods for disclosing truth should be doomed to failure. Critics who insist that talk of truth is jejune or that inquiry can only fathom the middle-sized part of nature must provide arguments to buttress their charges.

So too with the suggestion that the conception of an independent world is incoherent. Our commonplace thinking about ourselves and others embodies a natural epistemological attitude. Human beings form representations of the things around them, and those representations guide their behavior. Sometimes their representations adequately and accurately represent objects, facts, and events that onlookers can also identify; in other instances, possibly in many, people misrepresent the panoply of things around them. Contemporary ventures in cognitive psychology develop these commonplaces, but they already play a large role in our everyday lives, in parental guidance of child development, for example.

Thoughts of an independent world find their home in such mundane beginnings. Imagine yourself observing another person who is responding to an environment that you can also observe. It is clear that the imagined subject may form representations that are either accurate or inaccurate, that even if you were tempted to describe the contents of her representations as "true of her world" there would still be things independent of her of which those representations may not be correct. Similarly, you can reasonably suppose that she aspires to form accurate representations of the independent entities and that her success in responding to and shaping her environment reflects the extent to which she achieves her goal. Now you would be presumptuous to take yourself to be crucial to these judgments. She would interact with the same things and form the same representations whether or not you were observing her, and the relations between her accuracy and her successes would persist, unaffected. Moreover, like the subject, you could be viewed from the outside and assessed in a similar way. Recognizing all these points, it is easy for us to envisage a world of entities independent of *all* of us, a world we represent more or less accurately, and to conclude that what we identify as our successes signal the approximate correctness of some of our representations.

I have been trying to outline the homely motives for scientific realism. As we consider arguments that try to undermine our natural attitudes, I hope that the character of the position and its rationale will become ever clearer.

A first sobering thought is that realism is dogmatism by a nicer name. Claiming that the sciences tell us the truth in particular cases can easily sound like a refusal to countenance alternatives. So, for example, the realist suggestion that it's true that our atmosphere consists mainly of nitrogen and oxygen in a roughly 4:1 ratio elicits the complaint that "there are no absolutes." That charge is confused. To claim the truth of a statement is not to declare the certainty of our knowledge. Whatever hopes our predecessors may have had, contemporary views about human knowledge are saturated by the conviction that our beliefs about nature are fallible, that absolute certainty is not an option for us. When someone maintains the truth of the thesis about the composition of the atmosphere, he can consistently acknowledge the possibility that further inquiry might reveal it to be false. Indeed, we spend our lives proclaiming true, and acting upon, beliefs we recognize as vulnerable to the course of future experience. There is no snapping shut of our minds, no insulation against critical scrutiny, when we move from saying what we believe to declaring its truth.

Yet how can we maintain that what we believe is true when others disagree? Opponents of scientific realism are often impressed with the diversity of human beliefs. They suggest that one cannot account for the variation by appealing to the world, for the nature that the believers confront is the same. Provided we pay two dissenting parties the compliment of "natural rationality," we must suppose that both have done as well as possible. Both, then, must have attained "truth," and since their "truths" are different, we must reconceive "truth for a society" as what that society accepts, taking its "world" to be socially constructed, as the shadow of its "truth."

Influential though it is, the objection I have sketched is a muddle. People form beliefs about the world partly on the basis of their experience and partly by reasoning from that experience. Experience may be more or less wide, inferential skills more or less keen. When one can explain variation in belief by exhibiting differences in the range of experiences or the reliability of the reasoning, there is no dogmatism in claiming that one of the rivals is true and the alternative(s) false. Where such explanations fail, an undogmatic realist should suspend any attribution of truth.

In many instances, however, scientific realism makes judgments, and the judgments are castigated by opponents as chauvinistic, imperialistic, and insensitive. Who are we to assert the superiority of western science over the systems of thought that prevail in other regions of the world? That question should not be rhetorical. Consider the suggestion that western beliefs about the mechanisms of heredity are closer to the truth than those current among some culturally distinct group. Defense of the suggestion need not deny the "natural rationality" of members of this group. Instead, champions of genetics should point out that western scientists and their societies have had a greater interest in this topic, that our range of experience of hereditary systems is much broader, that we stand in a tradition in which substantial effort has been expended in building on the achievements of previous investigators, and so forth. Furthermore, we rightly appraise the tradition critically, and part of the critical attitude should lead us to inquire if the rival views, based on different experiences, provide grounds for revising or enriching our beliefs.

Claiming that particular sciences yield specified truths is thus compatible with recognizing that what we now think may require revision in the light of further inquiry, and with due respect for people and cultures who hold views diverging from our own. Dogmatic imperialism is not the inevitable result of rejecting the implausible idea that any system of belief adopted by any community has equal title to "truth." Yet my clarifications of scientific realism may seem to expose a new source of trouble in offering an overly optimistic view of human knowledge. Especially when we conceive nature as realists do, it may appear that any claims to have reliable procedures for arriving at truth about an independent world must be inflated. Realists offer us a vision of reality as something inaccessible.

Arguments that appeal to the inaccessibility of reality are the most powerful weapons in the antirealist arsenal. I shall begin with a version that is relatively simple and that elaborates the lines of criticism bruited in the discussion so far. An alternative to the realist explanation of human variation in belief proposes that all our opinions, even those that seem most closely linked to observation, are permeated with the concepts and categories in terms of which people make sense of the flux of experience. Different societies disagree because they apprehend nature through radically different conceptual frameworks—what they find in nature is what they bring to it, so that the only world that can meaningfully be discussed is "shaped" or "constructed" by the categories of some group. It's far from clear that this radical conclusion follows from the announced premises, but, in any event, the idea that our experience is determined by the categories we impose on it is surely too strong. Prominent examples from the history of inquiry drive home the moral that experience can sometimes violate prior expectations, prompting reconceptualization: Renaissance astronomers found new stars in the supposedly immutable heavens, Röntgen discovered X-rays, and, more recently, molecular biologists identified reverse transcriptase and enzymatic RNA.

Perceptual psychology furnishes the important insight that our representational states are the joint product of patterns of energy transmission from the world beyond the skin and the prior states of perceiving subjects. Antirealism thrives on combining that insight with an outdated model of perceptual processes. Seventeenth-century thinkers were held captive by a picture: the observer sits behind a screen, onto which images are projected. Add to that picture the idea that some of the features of the image stem from the character of the observer. How, then, can any observer ever have access to the Reality beyond the screen? How can the features that depend on the subject be disentangled from those that show things as they really are? There is no Archimedean point to which we can ascend to check which aspects of the image match reality—or whether or not any of them do. Once realists desert (as they must) the naïve account of perception that ignores the dependence on prior states of the perceiver, their conception of an independent reality makes skepticism unavoidable.

Perception is a process in which we perceive objects that are typically independent of us by *being in* (or *having*) representational states. We do not perceive by *perceiving* our states. Analogies with screens are profoundly misleading. Recall that the insight attributed to perceptual psychology was the product of experiments and observations designed to study the extent to which prior beliefs, concepts, and training influence the character of our perceptual states. Precisely such experiments and observations, conducted in physics, physiology, and psychology, illuminate the ways to answer any legitimate questions that the antirealist can pose. When the query, "How can we ever tell the properties of the real objects?" is transmuted into a (semi-)defensible form, we see it as calling for investigation of the causal factors that shape perceptual states. Rather than concluding that "the world" is shaped by our categories, realism points out that it is more accurate to say that our representations of the world are so shaped, that the source of the insight on which antirealist criticism draws is a collection of empirical inquiries, and that we can look to further inquiry for further enlightenment.

The riposte is obvious. When the status of the sciences as delivering truth about nature is at issue, it is question-begging to draw on the claims of particular sciences to vindicate the ability of perception to provide us with accurate knowledge of particular aspects of nature. Those convinced that the world about which we "know" is a world we have "constructed" are likely to protest that we should not appeal to the sciences to underwrite their own accuracy. Yet what exactly is the alternative? Unless the critic wants to abandon the opinions we all use daily in organizing our lives—including the doctrines about perception on which the antirealist's objection depends—it will be necessary to suspend the most global skeptical concerns. Philosophers sometimes invite us to throw away all our beliefs, start from scratch, and show that large swaths of accepted doctrine are true and that everyday methods of forming beliefs are reliable. That invitation should be rejected. Despite the many insights gained by thinkers in the Cartesian tradition, we now know that the enterprise of fundamental justification is hopeless, just as we know that it's impossible to trisect an angle with ruler and compass or to enumerate the real numbers.

Two different objections to realism must be carefully distinguished: one contends that, in light of what we think we know, there are grounds for holding that our perceptions don't accurately represent certain aspects of nature; the other demands that we demonstrate, without making any empirical assumptions, that our perceptions are accurate in the way we take them to be. The first point is serious, and leads directly to a scientifically informed scrutiny of our beliefs and our ways of generating them. The realism I commend, with its thorough commitment to human fallibility, is perennially concerned with recasting our beliefs in the light of what empirical sciences tell us about our relations with the world beyond our sense-organs. That process of revision already began in the early modern period, and it may be developed in new directions as the contemporary cognitive sciences uncover the complicated ways in which our representations are formed.[1] The second complaint embodies a sweeping demand for certainty, for *a priori* guarantees, that no position, realist or antirealist, can honor. Global skepticism is a great leveler, and screens, veils, cave walls, and social constructs are no more proof against it than the independent nature realists invoke.

Influential dismissals of the idea that the sciences can offer truths about a world independent of human cognition advance one or more of the arguments just reviewed: they allege a connection between realism and dogmatism, advert to the variations in human belief, or conjure a perceptual screen dividing the subject from the "real world." Although the reasoning is muddled, it gains credibility from links to more sophisticated lines of objection. Philosophical debates about realism turn on intricacies that do not surface in broader intellectual discussions, and gestures towards these more complex considerations lend cover to cruder attacks.

For the simple broadsides have subtle relatives. Some of them, the *empiricist* objections, try to drive a wedge between the claims of common sense about the observable parts of the world and scientific speculations about recondite entities. Others, the *constructivist* complaints, maintain that the realist's conceptions of truth and of an independent world are incoherent. In their quarrel with empiricists, realists issue a plea for democracy: both sides agree that statements in a privileged class (statements about observables) have truth values that depend on the way the (independent) world is, and the issue concerns the legitimacy of extending the privilege. Constructivists, by contrast, typically don't want to oppose the democratic tendencies, but take a different view of the character of the privilege. Their concerns are often seen as "deeper," but, in understanding the merits—and limitations—of realism, we do best to start with the realism–empiricism debate.

One important line of empiricist argument applies the point about variation in beliefs to the history of inquiry. Many people are impressed by the fact that current sciences are extremely successful in predicting future events and in enabling us to intervene in our environments. If our appreciation of this success is

1. I am grateful to Patricia Kitcher for helpful advice on this score. Within psychology, for example, constructivist approaches to representational states might undermine our ascription of familiar properties to familiar objects. The possibility is compatible with the modest realism I defend, and that form of constructivism is different from the philosophical doctrines I oppose. The difference can be seen clearly by reminding ourselves that, on the psychological account, it's the *representation* not the *world* that is constructed. As the discussion of chapter 4 will make clear, there are strands in the psychological forms of constructivism with which I am very much in sympathy.

to justify us in believing that the statements from which our predictions and interventions flow are true, then it will be important to suppose that our predecessors, whose theories we now reject, could not have made a similar case for their own views. When we inspect the historical record of inquiry, we find, however, that the past is littered with discarded theories whose proponents took them to be successful in just the ways we view our contemporary sciences. Inferring truth from our impression of success is thus an erroneous dogmatism.

Success is naturally taken to betoken truth: the predictions of particle physicists and the organisms manufactured by molecular biologists inspire us to ask how we could obtain agreement to ten or so decimal places or breed mice with tissue characteristics made-to-order unless our theoretical claims were at least approximately true. The examples illustrate the type of success realists have in mind when they wax eloquent about contemporary science. A theory is not successful for a group of people simply in virtue of the fact that members of that group can use it to attain their goals. Groups can set their sights low and obtain facile "successes"; conversely, successes can be available, whether or not they harmonize with the contingent goals of some cluster of people. On the realist conception, a science is successful when it provides the means to hypothetical ends, ends which are numerous, diverse, and individually difficult to achieve. Appraisals of success are fallible, for we may be misled about whether our predictions and interventions meet these criteria: to cite just one possibility, the task of generating so exact a prediction may not be as exacting as we think.

Judged by these standards, does the history of the sciences reveal numerous outmoded theories that were genuinely successful, or that were reasonably viewed as successful by their protagonists? Historically-minded antirealists are fond of drawing up long lists, in which they hail such defunct doctrines as the humoral theory of medicine, phlogistonian chemistry, and theories of spontaneous generation as successful.[2] Examples like these invite a direct response. None of the theories in question was able to yield predictions and interventions that were numerous, diverse, and hard to achieve, and none could reasonably have been taken to do so.

Yet even after the inflated claims about successful-but-false theories have been rejected, realists must confront more troublesome examples. Atomic chemistry yielded vast numbers of empirical successes throughout the latter half of the nineteenth century, even though its understanding of atoms was radically incorrect. More striking yet, the wave theory of light, from Fresnel to Maxwell and beyond, issued predictions that nobody had expected. Despite its commitment to the existence of an all-pervasive ether, the theory generated some precise and counterintuitive expectations about what would be observed

2. See, for example, Larry Laudan, "A Confutation of Convergent Realism," *Philosophy of Science*, 48, 1981, 19–49.

under special conditions. As Poisson, a staunch opponent of the theory, pointed out, Fresnel's mathematics of wave propagation yielded the conclusion that, when light shines on one side of a small disc, there should be a bright spot in the center of the disc's shadow. Much to his surprise, experimentation demonstrated the existence of the bright spot—which, ironically, is known as the "Poisson bright spot," bearing the name of the man who doubted it. If the inference from success to truth is reliable, then successful theories ought to be approximately true, and, in particular, the putative unobservables to which they appeal ought to exist; but, contra Fresnel, there is no ether; hence the realist's preferred inference is unreliable.

Reliability doesn't require that the inference lead to true conclusions in every case, and one response might be to insist that the example of Fresnel is exceptional. Realists can do better than special pleading, however. Why was Fresnel's account of light so successful, given that his beliefs in the ether and his identification of light as waves in the ether (jostlings of ether molecules) were wrong? Contemporary optics textbooks reproduce Fresnel's mathematics of wave propagation, and they derive the phenomena reported in his experiments in the way he did, by showing how the mathematical formulations yield patterns of interference and diffraction (including the pattern that gives rise to the Poisson bright spot). Fresnel was not completely right about light, but it would be folly to leap to the conclusion that his theory is completely wrong. Nor should we think of that theory as some indissoluble unit in which the infection of error spreads from hypotheses about the ether to invade the whole. His mistaken belief that all wave motion requires a medium in which the waves are propagated (the belief that led him to posit the ether) is irrelevant to his mathematical explanations and predictions. His successes stem from his use of parts of the theory we continue to honor as approximately correct.

The case of Fresnel points to a general strategy for addressing pessimistic reflections on the history of science. Instead of thinking about the virtues and vices of whole theories, condemning as false those that are not (by our lights) entirely true, we should separate those parts of past doctrine that are put to work in prediction and intervention, the "working posits" of the theories, from the "idle wheels." The practice of inferring truth from success should be understood as one in which we support belief in the working posits of our theories. Idle wheels gain no credit from the theory's successes, and, by the same token, historical examples in which critics point out that we have discarded some of the idle wheels of the past leave realism unscathed.

Hindsight is well known to be wonderful in clarifying vision. Contemporary thinkers find it much easier to pick out the working posits and the idle wheels of the wave theory of light. One might even wonder if it's possible for practitioners to draw this distinction, so that, when we turn to our own theories, we'll find ourselves unable to identify those parts for which we can justifiably claim truth on the basis of success. Yet, even though Fresnel *didn't* make the distinction between those parts of his theory that were implicated in his successes and those that were not, it's far from obvious that he *couldn't* have done so. Moreover, even if we concede, pessimistically, that the working parts of a theory are intertwined with theoretical excresences that only later analysts can cut away, the fallibilist realism I have been advocating has an obvious proposal. Our predicament is very like that of the historian who writes a detailed narrative of a complex sequence of events: each sentence is well-grounded in archival research, so that belief in the truth of the individual parts of the story is justified; yet, as any reflective historian knows, there may be missing perspectives that need to be supplied by others, so that those who come later will disentangle truth from falsehood in ways that are presently unspecifiable; for the moment, the historian reasonably expresses confidence about each component of the narrative, while admitting that it's overwhelmingly probable that there's a mistake (or a faulty conceptualization) somewhere; an analogous attitude is expressed in a modest realism about the sciences.

Defending the strategy of inferring truth from success against a bevy of historical examples is only the first step in turning back empiricist challenges to realism. For thoughtful empiricists will deny that the burden of proof lies with their showing that the strategy is unreliable. Rather, they will point out, we need to be given some reason to think it might be reliable. Here, it seems, realism encounters a fundamental problem. The only claims we can check directly are statements about observable things and events. If we apply a method of justification to go beyond such claims, then we should have some assurance that the method is likely to lead us to correct conclusions. But we can only check methods by looking at the conclusions they generate about observable things and events. It follows that we have no basis for trusting methods insofar as they yield conclusions about unobservable things and events, and that the appropriate stance to take to those parts of our theories that talk about unobservable entities is an agnostic one.

From the empiricist perspective, realists typically do two dubious things at once, make claims about unobservable entities and apply methods that yield conclusions about those entities. The application of the methods would only justify the conclusions if the methods could be justified in that type of application—and that would seem to require an independent way of validating the claims about the unobservable entities. So characterized, realism appears to face a hopeless predicament. Yet historical reflections provide glimmers of hope. For there have been several occasions on which scientists have encountered similar challenges—and have overcome them.

In 1610, Galileo Galilei faced a difficult problem. He claimed to have an instrument that would disclose wonders, novel things in the heavens, beyond the sphere of the moon, where, according to Aristotelian orthodoxy, no changes were allowed. His instrument was new, too, and his adversaries lost no time in protesting that it was unreliable and its deliverances about the (immutable) heavens quite erroneous. How then could he simultaneously argue for the correctness of his announced discoveries and the trustworthiness of the telescope?

Galileo reacted to this problem in a straightforward way, by showing that the telescope would deliver conclusions that could be verified using methods his contemporaries, including his critics, would accept. Looking toward a distant building, they could use the telescope to read the letters carved into the stone façade, then verify their judgments by going closer to look. Pointing the telescope at ships bound for port, they could detect a sailor with a parrot on his shoulder, pick out a broken spar, and survey some of the cargo, predictions that could be confirmed when the ship arrived. Even Galileo's most committed opponents were forced to admit the power of these demonstrations. "Below it works wonders," conceded the unscrupulous Martin Horky, in the thick of a fierce denunciation of the astronomical uses of the telescope.[3]

What Galileo showed initially, before the telescope was widely distributed, was that his instrument could reliably disclose terrestrial phenomena in various parts of Northern Italy. Even before anyone had checked, there was little plausibility to the thought that it would work in Venice but not in Amsterdam or in London, for Holland and England were not distinguished from Italy by any property pertinent to the functioning of an optical instrument; nobody suggested, for example, that the telescope would only reveal distant objects in countries that were officially Catholic. By contrast, there was a well-recognized distinction between the sublunary world and the terrestrial sphere, and critics of the telescope, like Horky and his less malicious fellow-travellers, denied the legitimacy of extrapolating from reliability on earth to reliability in the heavens. Galileo's central problem was to make the celestial–terrestrial distinction appear as irrelevant as the difference between London (or Amsterdam) and Venice.

He solved the problem by using two kinds of arguments. The first exploits the fuzziness of the boundary between the directly observable and the unobservable, emphasizing the continuity between what is at the limits of unaided human observation and what is clearly discernible through the telescope. Astronomical observers make out individual stars in a constellation with more or less difficulty, and those equipped with especially acute vision—tested, of course, in sublunary situations—just make out a pattern of distinguishable stars where others see a blur; armed with the telescope, both types of observers see the pat-

3. See Albert van Helden, *The Invention of the Telescope* (Chicago: University of Chicago Press, 1985), and van Helden, ed., *Galileo's Siderius Nuncius* (Chicago: University of Chicago Press, 1989). As a number of historians have made very clear, Horky's fervent opposition to Galileo was viewed by his contemporaries as dishonest.

tern clearly and without strain. The second style of argument undercuts the significance of the celestial–terrestrial distinction by cataloguing changes in the supposedly immutable heavenly sphere. Naked-eye observations of new stars and comets reveal that there is no basis for thinking of the heavens as importantly different from things "beneath the moon." Combining these two arguments with his ability to distribute telescopes that would generate an increasingly more consistent set of astronomical observations, Galileo was able to convince his peers that there was no more basis for thinking the instrument was unreliable in the heavens than for believing it inept in some as yet untried part of the earth.

Methods of justification, like Galileo's telescope, can only be validated by examining the conclusions about observables to which they lead. It does not follow that the only conclusions licensed by those methods are conclusions about observables—any more than Galileo's demonstrations on buildings and ships only show that the telescope is reliable in Venice. We need to consider whether there are good reasons for distinguishing a method's usage in its application to observables from its usage in application to unobservables.

Why do we believe that statements used to make successful predictions and interventions are likely to be true? Reacting to the earlier complaint that the history of science reveals the unreliability of such inferences, realists might suggest that our confidence rests on serious analysis of past ventures in inquiry. To do so would be (at best) to engage in a polite fiction about sustained archival investigations. There is a much more mundane source of the belief that success indicates truth.

People find themselves in all sorts of everyday situations in which objects are *temporarily* inaccessible, or are inaccessible to only some of the parties. Detectives infer the identities of criminals by constructing predictively successful stories about the crime, bridge players make bold contracts by arriving at predictively successful views about the distribution of the cards, and in both instances the conclusions they reach can sometimes be verified subsequently. We readily envisage an idealized type of situation, perhaps most perfectly realized in some parlor games, in which the "success to truth" rule is tested and confirmed:

1. There is a class of people who are trying to understand and predict aspects of the behavior of a system. These people have no direct access to the entities causally responsible for that behavior.
2. Each person in the class has a set of views about the underlying entities, either through information given in advance or through the formation of individual opinion.
3. The predictive practices of the people in the class are more or less successful. Either they themselves, or observers, can eventually identify the underlying entities, assess the correctness of the views used to generate

predictions, and discover the correlation between accuracy and predictive success.

Very probably, nobody has ever designed and run carefully controlled studies to find out just what occurs in such stylized situations, but we all have a large body of experience of cases that approximately satisfy these features. On the basis of these experiences, we judge that the correlation between success and accuracy is high. More exactly, we come to believe that people usually only manage to achieve systematic success in prediction when their views about the underlying entities are roughly right.

How does this belief bear on the dispute between realists and empiricists? Realists see the everyday phenomena to which I've alluded as showing the reliability of a rule licensing us to conclude that successful practices indicate roughly correct ideas about the underlying entities described by the statements that do the generative work, whether or not the entities in question are observable. Empiricists suppose that the same phenomena only justify the rule subject to the constraint that the entities are observable. At just this point, realists can invoke the Galilean strategy. Empiricists must hold that the world is so adjusted that a perfectly good method turns unreliable when it is applied below the threshold of our (contingent) powers of observation. Although they frequently urge the epistemological modesty of not committing oneself to judgments about unobservables, from the perspective of the Galilean strategy, maintaining that inferring truth from success breaks down at a point fixed by some idiosyncracies of our species looks like the heights of metaphysical hubris. Why should the reliability of the rule differ when we can no longer assess the consequences?

Galileo invited his contemporaries to consider whether traditional boundaries mark significant differences. We should do the same. The distinction between the observable and the unobservable is not sharp. Just as Galileo exploited the fuzziness of the boundary, emphasizing the continuities of naked-eye and telescopic observation, we can ask pointed questions. Suppose there is just one person, Hawkeye, who can detect entities smaller than those visible to anyone else. Does this mean that the strategy of inferring truth from success has a slightly wider legitimate application just because of Hawkeye's (contingent) existence? Without Hawkeye, the pertinent entities could only be disclosed to human beings by interposing pieces of glass. But just what difference does that make? Why should it be taken as a warning against applying our everyday inferential strategy? As we recognize the contingencies of human perception, we realize that empiricists rely on a distinction between observable and unobservable strikingly akin to the distinction between telescopic observation in Venice and telescopic observation in London—or to the distinction between the celestial and the sublunary after Galileo's critique of it.

Our knowledge of the diversity of situations in which people form beliefs about temporarily inaccessible objects provides a *positive* view of the conditions under which the inference of truth from success will be reliable. Criteria of predictive or interventional success can vary enormously from case to case. Sometimes, even though there are many possible outcomes, they divide into a small number of types and success only requires identifying the appropriate type; on other occasions, people must make a fine-grained identification to be successful. Similarly, there are tasks that are much more error-tolerant than others, in which one only has to have a coarse idea about what is going on. Finally and familiarly, people with faulty views can make compensatory errors, while those with correct beliefs may misapply them. All these features need to be understood in considering how everyday experience supports the inference from success to truth.

The commonplace that the ignorant can be lucky and the learned unfortunate is readily accommodated by requiring that success be systematic. Although inaccurate beliefs can sometimes combine to yield happy conclusions, we know that it's improbable that they should do so, and the chance diminishes with the range of situations across which the beliefs generate predictions. Far more dangerous is the suggestion that success is an artifact of the error-tolerance of the prediction task. If all that is needed to generate the right prediction is to have one true belief about one of the entities involved, then success doesn't redound to the credit of the detailed story told about the rest. The right way to raise a challenge to the realist's inferential strategy is to maintain that we don't know whether the practitioners of successful sciences are confronting error-tolerant problems or not.

Ironically, this challenge fails to support the empiricist theses about the domain in which such inferences are reliable. Doubts about whether or not the situation we're confronting is or is not error-tolerant can arise both when the underlying entities are observable and when they are unobservable. Thus the division between cases in which the method is reliable and those in which it isn't cannot *coincide* with the distinction between conclusions about observables and conclusions about unobservables. Nor can it legitimately be claimed that the cases in which the method works properly are a subclass of the instances in which the underlying entities are observable. There is no reason to think that error-intolerant situations are just those in which underlying entities are observable, or to hold that we can only justifiably claim that a situation is error-intolerant when we know that the entities that figure in the predictive practice would be observable. As always, we should apply the Galilean strategy and ask what our experience of situations we can actually check teaches us about the pertinent distinction (in this case, error-tolerance versus error-intolerance).

Here the distinction between fine-grained and coarse-grained prediction proves useful, for if we have any basis for a correlation between a detectable feature of the situation and the error-intolerance of the predictive task, it is that the ability to make fine-grained predictions indicates that the task is unlikely to be error-tolerant. Hence, the best judgment we can make about the reliability of the "success-to-truth" rule is that when the predictive success is both systematic and fine-grained the inference is most likely to be reliable—as when we make predictions that are accurate to ten places of decimals or design complicated organisms to exhibit combinations of properties quite unprecedented in nature. But, of course, these properties of the predictive practice are orthogonal to the division between observables and unobservables.

So, maybe we can claim a victory for realism in the debate with empiricists, but can it turn back the "deep" objections? The serious challenge is the constructivist complaint that talk of an "independent world" is incoherent. How can we extrapolate from everyday practices to metaphysical conclusions? Just this way of posing the question seems to me to be misguided. Antirealism thrives on supposing that there is an enormous gulf between the realist's claims and everyday ideas and judgments, on accusing realists of importing unnecessary metaphysics.

Realism, I have suggested, begins with reflection on homely examples. Our everyday understanding of people and their actions provides us with a concrete example of the view that others represent objects and states that are independent of them. Imagine you are watching someone using a map to navigate unfamiliar terrain for the first time; for example, suppose the subject is a newcomer to London and she is using the standard map of the Underground.

Central to our ordinary explanation of what the subject does is the idea that she represents objects that would exist even if she were not present. We take the dots on the map as corresponding to things we can pick out in her environment (underground stations), and we think that when she reads the map there are elements of her representation that also correspond to those things. Our basis for making this attribution is the distinctive role dot and station play in coordinating her behavior. "This is Charing Cross," she says, "I must change here"—and so she does. A few of her friends act differently, so that, for them, dot and station play a different role, and we regard them as having a different correspondence between mental counterpart of dot and station—more colloquially, we describe them as misreading the map. Finally, we don't think of ourselves as necessary to her performance or as altering its causal structure. The correspondence we recognize between her mental states and the world is disclosed through her behavior, not created by our observation and explanation of it. That correspondence is grounded in the ways in which mental elements figure in the production of behavior with respect to certain kinds of surrounding objects, and it would remain as it is even if we were not standing by to observe.

Constructivists question the coherence of the idea of an independent world. I propose that we understand the notion of independence by thinking about situations in which the independence can be observed (as we recognize that the stations remain while our passenger comes and goes), but we can use that notion more broadly to encompass situations in which no observer is present. Terms like 'part of,' 'little,' and 'person' are initially introduced in situations in which the things that serve as exemplars are macroscopic objects. Yet they do not only apply to observable things: we can explain the (nineteenth-century) concept of atom by declaring that atoms are those parts of matter that themselves have no parts, and most children readily understand stories about "people too little to see."[4] So too the notion of 'independence' is initially introduced in situations where the independent objects are—or can be imagined to be—observed. I suggest that we can understand talk of objects independent of all cognition by extrapolating in just the ways we do with other concepts. Our purchase on the idea that some objects are independent of some of us (although observed by others) suffices to make intelligible the thought that some objects are independent of all of us, that they would have existed even if there had been no humans (or other sapient creatures), even though, had that been so, there would have been no observation of them or thought about them.

Can we have direct access to such objects? Of course. This can be seen from the example of our subject navigating her way. Although the objects she confronts and with which she coordinates her behavior exist independently of her, her access to them is perfectly direct: while waiting on the platform at the station, she may have indirect access to the approaching train by observing the monitor that announces its arrival, but, when it actually comes, there is no plausible sense of 'direct' in which her access to it is less than direct. Realists should however agree with constructivists that this direct access is mediated by concepts and categories; as we saw earlier, we perceive independent objects by being in states whose features are partly caused by the characteristics of the subject. There is no Archimedean point from which any of us can look down on an unconceptualized world and inspect the relations between the independent objects it contains and our representational states.

Constructivists think the Archimedean point is necessary if realists are ever to know anything about independent objects—or even to use language that refers to such objects. Consider the latter point first. The charge is that a sign ("Charing Cross" or its mental counterpart) cannot be determinately linked to a cognition-independent object (the station Charing Cross) unless there is some special place from which the great linguistic inaugurator can point to both at once. Nobody should believe in the inaugurator or in the privileged place he is supposed to occupy. But why should this prove troublesome? There is an obvi-

4. These important points, and the particular example, stem from Hilary Putnam, "What Theories Are Not," in his *Mathematics, Matter, and Method: Philosophical Papers, Volume I* (Cambridge: Cambridge University Press: 1975), 215–227.

ous realist account of the ways in which links between language and the world are forged: connections between signs and objects are constituted by the pattern of causal relations involving objects, representations, and behavior; what makes the terms we use refer to the entities they do is a matter of the situations to which we respond in using them and the ways in which our usage of the terms guides our behavior.

Consider, once again, our imaginary subject traveling under London. We take her terms to refer to various lines and stations because of the circumstances in which we see her produce them. Alternative proposals about language–world connections would not enable us to understand her behavior. Is this sufficient? Constructivists may protest that all that has been done is to demonstrate that, within a particular framework for conceptualizing the world, we can use the ordinary notion of reference and suppose commonplace links between words and objects. But that is taken to work only because someone—the observer—looks on and recognizes signs, objects, and their systematic relationships. Since it has already been agreed that the analogous idea for identifying a connection between signs and mind-independent objects (the great inaugurator, the Archimedean point) is absurd, the employment of ordinary referential locutions and their ordinary role in explaining behavior cannot be extended to the context in which it is assumed that objects are independent of all of us and of all cognition.

To make further progress it is necessary to recall an obvious point. We believe that there is a pattern of causal relationships linking the objects to which the subject responds, the linguistic and mental signs she employs, and the ways in which she behaves. One day, we hope, cognitive psychology and linguistics will combine to tell us more about this pattern. The crucial point, however, is that, whatever this pattern is, we don't think that it depends on the presence of the observer. Even if we were not around to watch her performance, she would still go through the same psychological states and perform the same actions. Thus the relation of reference may be *discovered* by the outside observer, but it is not *set up* by the observer. Hence we can envisage that that relation should obtain for each of us independently of the presence of any other observer, and for *all* of us without any great inaugurator to occupy the absurd Archimedean point.

So far, I have addressed the complaint that it would be impossible for us to talk about the world as conceived by realists, but there is a parallel concern about the possibility of knowledge. To the realist proposal that we assess our representations by our practical successes in negotiating the environment, constructivists counter that this is all very well so long as we refrain from thinking that success is any sign that we are accurately representing the properties of mind-independent objects. The crucial objection is this: whatever reasons we have to think success betokens truth in our representations of objects of experience do not apply when we are considering truth about mind-independent objects; for, once it's conceded we lack an Archimedean point that would provide access to these objects, there's no basis for checking the correlation between success and fathoming the properties of such objects.

We encountered an analogous argument in the realist–empiricist debate. Its centerpiece consisted in the denial of any possibility of checking methods in their generation of conclusions about unobservables. Realists reply by asking if there's any reason to suppose that a method whose applications can be verified in one domain breaks down when we move to a broader domain. Pressing the Galilean strategy, they question the relevance of our contingent powers of observation to the connections we observe between success and accuracy. In the present instance, however, constructivists have an apparent means of resisting parallel reasoning. Can't they invoke a distinction more durable than the celestial–terrestrial distinction to block the attempted extrapolation, the distinction between objects-as-experienced and objects-in-themselves?

Surely this is an exercise in dubious metaphysics. Realists do not see any distinction between objects, between Charing-Cross-as-experienced and Charing-Cross-in-itself.[5] There is just Charing Cross, sometimes experienced, sometimes not, and, when we experience it we are able to recognize some of its properties, and, we believe, to do so accurately. It's simply a mistake to think of our different relations to the same entity as bringing into being different objects. Hence the response is even simpler than the counterpart reply to the empiricist, for there's no distinction of entities about which we draw conclusions on the basis of our successes—the objects we claim to represent accurately are not mysterious things-in-themselves but, in many cases, the things with which we interact all the time.

Perhaps this simple response will seem too blunt. We do well to return to an everyday example in which the explanation of success can readily be assessed. Imagine, once again, our London traveler and suppose that, unbeknownst to her, a band of Sinn Fein sympathizers has decided to subvert metropolitan life in a relatively non-violent fashion: each night they replace the signs announcing stations that are visible from the train with incorrect ones. Initially, our subject, like many others, is confused. She gets off at the wrong stations and finds herself in the wrong parts of London. After a time, she comes to rely on the Underground map and to count carefully the stations between her starting point and her intended destination. The procedure works, and she goes where she

5. This is not to deny that our experience of Charing Cross comes about by a complex process that cognitive psychology might explore, nor to suppose that, in light of those psychological explorations, we might come to revise judgments about exactly where we are successful, and why. Rather the point I'm insisting on is that our experiential state doesn't relate us to some special object, Charing-Cross-as-experienced, and fail to relate us to another special object, Charing-Cross-in-itself. The experiential state is a representation of Charing Cross; the success of the activities to which that state contributes suggests the representation is accurate in important respects; but, as I've conceded, scientific research might lead us to qualify that judgment.

wants to go. We, the observers, understand her success in terms of the accuracy of the map, recognizing her as justified in overriding the judgments she would normally make by casual observation from the train window. If the map were not accurate then she would not do well: as always, we rely on our common experience of likely success rates with accurate and inaccurate representations. Perhaps we even do an experiment, supplying some passengers with inaccurate maps, and confirm the correlation between accuracy and success in this particular case.

Our presence is not necessary for the subject to be successful. She would do the same things in our absence, and the explanation of her success would be the same. Of course, we can never verify the connection between success and accuracy of representations on those occasions when no observer is present to inspect the objects represented—the pertinent entities cannot simultaneously be both observed and unobserved!—but once we hold firmly to the idea that the presence of observers doesn't make a difference in the important respects, there should be no temptation to believe that the reliability of the inference from success to truth is undermined on these occasions.

The next step is to generalize, claiming that the success that people collectively enjoy in predicting the behavior of objects that exist independently of all of us and in adjusting our actions to them indicates that our most successful ways of representing the world are approximately correct. It's crucial to see clearly what is going on at this stage. Realists don't suppose that because the inference from success to truth works well for a certain kind of entity—things-as-they-appear-to-us—that inference will also be reliable in connection with a different kind of entity—things-as-they-are-in-themselves. That would invite the charge that there's a crucial difference that blocks the right to project from the homely examples to the Grand Metaphysical Conclusions. Instead, the generalization works by supposing that there are two kinds of situations, those in which the properties of things are detected by observers and those in which they are not. Because realists see no causally relevant difference between these two kinds of situations, they maintain that the inference from success to accuracy is reliable in the second type provided that it is reliable in the first.

If my arguments are correct, then the intellectual fashion of deriding simple scientific realism is unjustified. The sciences sometimes deliver the truth about a world independent of human cognition, and they inform us about constituents of that world that are remote from human observation. As we shall see later, this minimal realist claim does not commit us to grander doctrines that are often announced on the banners of realism. Our immediate task, however, will be to understand an ideal of objectivity for our knowledge of the world as we find it.

THREE

The Ideal of Objectivity

When a handful of distinguished gentlemen came together in post-Restoration England to set up the Royal Society, they agreed that membership should be open only to the better sort. Allowing tradesmen and artisans to join the collective search for truth seemed too dangerous to be tolerated, for, after all, the worldly interests of such people might corrupt their decisions about what counted as genuine knowledge. Over three centuries later, we have come to understand that even gentlemen, "free and unconfin'd," with all the advantages of the best educations, can sometimes fudge the data, report experiments that were never done, or be led to advocate unsupported conjectures because of their personal or political passions. Although we have learned that adopting the detached scientific stance, dedicating ourselves to the pursuit of truth and ignoring mundane distractions, is psychologically far more complex than the first members of the Royal Society imagined, many contemporary scholars continue to honor the ideal of objectivity. Indeed, that ideal is central to one of the images I presented in the first chapter.

Yet the ideal cannot be taken for granted. There are powerful arguments that threaten it, and the defense of simple scientific realism is incomplete until those arguments have been reviewed and analyzed. Enthusiastic about their new experimental methods, the scientists of the seventeenth century believed they had only to listen with open ears and open minds and Nature would whisper sweet truths to them. Their successors, pursuing difficult projects in the various sciences and reflecting on their frequent controversies, came to see that the standards for warranted judgment were much more intricate than it initially appeared. To this day, nobody knows how to articulate clearly and precisely an account of objective evidence that can be applied to the range of complex scientific debates—from the seventeenth-century disputes about mechanics and the propagation of light through eighteenth- and nineteenth-century controversies about chemical combination, the age of the earth, the history of life, and the structure of matter, to our current discussions of cosmological theories and the rival importance of nature and nurture. We lack an analysis that will reveal exactly which claims are justified (to what degree) by the evidence available at various stages. So there arises the suspicion that our lack of any such account reflects the impossibility of the project. Perhaps there are no standards for "objectively warranted judgment." Perhaps, in the end, all our opinions about the world are inevitably pervaded by our values and prejudices, and are none the worse for it.

The last chapter tacitly set suspicion to one side in its proposal that we could "reasonably" appraise scientific claims in terms of their successes. There are two obvious objections. The first would claim that appraisals of success are inevitably dependent on our goals and values: whether a scientific theory appears successful depends on whether it advances our efforts to attain the ends that matter to us. My specification of the idea of success was intended to forestall that criticism. Whether or not we value predictions and interventions of a particular type, we can recognize that a theory generates those predictions and interventions. So it is at least possible to take the first step and to identify the predictions and interventions that flow from a theory without the intrusion of value judgments. Can we go further? Are there objective ways of resolving situations in which alternative theories are associated with successes that are partially overlapping, partly disjoint? Can we decide among theories which yield exactly the same predictions and interventions? These questions formulate the second objection. Answering them will be the project of this chapter.

Much philosophical ink has been spilt around the second question. There is a doctrine with a resonant name—The Underdetermination of Theory by Evidence (the underdetermination thesis, for short)—which has passed from philosophy into more general discussions of science. In its simplest form, the thesis claims there are alternative theories which are not simply equally well supported by any evidence we have but which would continue to be equally well supported given any amount of evidence we could ever collect. Scientists are thoroughly familiar with the predicament noted in the first half of the thesis. Part of the routine character of their work consists in recognizing that different hypotheses are equally justified in light of the body of findings so far assembled and in devising experiments or observations that will enable them to resolve the issue. *Transient* underdetermination is familiar and unthreatening. The underdetermination thesis envisages situations in which the ordinary remedies break down. *Permanent* underdetermination occurs when, for any further results that might be garnered, there is always a way to extend each of the rivals to obtain theories which continue to be equally well supported.

Nobody should be much perturbed so far. It's hardly news that there might be two cosmological views that agreed exactly on the details of those parts of the universe to which we can have access but which differed with respect to regions of space-time with which human beings can never have any causal interaction. There may be some questions about nature we recognize as genuine issues but which we can never settle. The underdetermination thesis obtains its bite when permanent underdetermination is taken to be rampant. The global underdetermination thesis, which contends that all (or virtually all) instances of what scientists treat as transient underdetermination are, when properly understood, examples of permanent underdetermination, poses a genuine threat to the ideal of objectivity.

For in the mundane cases of transient underdetermination scientists do resolve the dispute between alternative hypotheses, opting for one and rejecting its rival(s). According to the global underdetermination thesis, there is a way of developing the rejected rival(s) to obtain a theory (theories) that would be just as well supported by the new evidence—the evidence that allegedly puts an end to debate—as the doctrine that is actually accepted. Scientists thus make choices when there is no evidential basis for doing so. How do they break the ties? At this point, critics of the ideal of objectivity often insert their own psychosocial explanations. Since there are no objective standards for judging the victorious hypothesis to be superior, the decision in its favor must be based on values: scientists (tacitly or explicitly) arrive at their verdict by considering what fits best with their view of the good or the beautiful or what will bring them happiness.[1]

The ideal of objectivity turns out to be a fiction. We thought we could distinguish (say) the researchers for Big Tobacco whose conditions of employment made certain genetic hypotheses highly attractive to them, or the craniometers whose skull measurements supported their prior views about the intellectual capacities of different races, from all those investigators who make their judgments purely in light of the evidence. We were wrong. In every instance, the evidence is incapable of showing that one hypothesis is superior to its rival(s), and, if the value judgments of an Einstein or a bench scientist are less overt than those whose "objectivity" has been impugned, they are, nonetheless, present. All scientists believe what they want to believe. Truth has little or nothing to do with it.

There is surely something puzzling about this dismal story. Crucial to the ar-

1. A classic version of this view is the reconstruction of the Boyle–Hobbes debate in Steven Shapin and Simon Schaffer, *Leviathan and the Air-Pump* (Princeton: Princeton University Press, 1985).

gument was the idea of global permanent underdetermination, and yet, at the concluding stage, we are told that scientists' values can work a magic that mere evidence cannot manage. Allegedly, for any two distinct bodies of doctrine, we can continue to maintain either in a way that will render it as well supported by the evidence as its rival. Let's play along, and suppose that the dispute in question is between a theory that allows for the inheritance of acquired characteristics and one that denies this. The scientific community claims to have resolved the dispute by particular experiments, but, according to the underdetermination thesis, this is not so, and there's a way to extend the rejected doctrine to a position that is just as well supported as the resultant orthodoxy. How did the pseudo-resolution occur? Assume that the preferred theory seemed to the scientists to be better at promoting their political goals. So there is a widespread belief maintaining that one of the rivals is more politically effective than the other. But now let us return to the global underdetermination thesis and start with a new pair of doctrines: one of these is the victorious theory coupled with the widespread belief that this theory is more effective at promoting the political goals; the other is the defeated doctrine supplemented with a contrary belief, to wit that *this* doctrine is politically more effective. If we are serious about the global underdetermination thesis, then there ought to be ways of extending our new rivals to make them equally well supported whatever the evidence. Hence in the original decision situation, the resolution couldn't have been achieved in the way that the critic of objectivity suggests. Not only couldn't the evidence have broken the tie, but the appeal to political values couldn't have done it either.

Perhaps this will seem like trickery. Instead, I submit, we should view it as drawing out the implications of the global underdetermination thesis. If *any* doctrine, no matter how ambitious, can be sustained in the march of inquiry, then claims about the impact of doctrines on our goals can be absorbed within the rivals we consider, and these, too, will prove vulnerable to the thesis. An odd, and unnoticed, feature of the influential ways of appealing to underdetermination to "expose the myth of objectivity" is that they simultaneously deny the possibility of rational standards for assessing evidence and assume the possibility of treating those who make scientific decisions as rational agents who are trying to promote "nonscientific values." The argument of the last paragraph suggests that one cannot have it both ways.

Tu quoques bring cold comfort. Psychosocial explanations may be undermined by the arguments that led people to assert their inevitability, but the ideal of objectivity still suffers if we adopt the global underdetermination thesis. For, if that thesis is true, there is no evidential basis for most scientific decisions, and the acceptance of particular hypotheses must reflect either the dogmatism of rejecting articulated rivals or a failure of imagination in developing the rivals (or possibly both). So it's important to examine the credentials of the idea of permanent underdetermination.

Where does that idea come from? Two important episodes in the history of physics have had an important philosophical legacy. One is the sequence of efforts to determine the motion of the solar system with respect to absolute space. The other is the development of two different versions of quantum mechanics by Schrödinger and Heisenberg in the 1920s.

In formulating his dynamics, Newton makes an important distinction between relative motions and true motions and frames his laws in application to true motions. Our observations of motion are always relative to a set of objects (or a frame of reference) which we treat as being at rest. What, then, can we mean by talking of "true" or absolute motion? Newton answers this question by introducing a special frame of reference, absolute space, and supposing that true motions are motions relative to absolute space. Natural philosophers, then and now, quickly see that it is going to be very hard to tell whether or not some systems are in true motion or are truly at rest. Newton thought he could show that the frame of reference defined by the center of mass of the solar system was one to which his dynamical laws could be applied, but it's compatible with that to suppose either that the frame is truly at rest or that it moves with any chosen constant linear velocity. So we can consider an infinite collection of rival theories, sharing the common core of Newtonian dynamics and the theory of gravitation, distinguished from one another by the values of a parameter that measures the motion with respect to absolute space. Each of these theories will predict exactly the same checkable consequences about motion—for, recall, we only observe relative motions. Many scholars who have reflected on the example have concluded that all the theories are equally well supported by any possible evidence, since any prediction that redounds to the credit of one will give an equal boost to any of the others.

How might we try to discriminate among this vast collection of rivals? One thought is that it's easy to take a myopic view of the evidence available to us. Perhaps we are not restricted to testing these theories by observing motions, perhaps in the advance of inquiry Newtonian dynamics might be integrated with some future theory that would make the value of the velocity with respect to absolute space pertinent to something that could be observed in a different way. Let us be fanciful. Suppose scientists later come to accept the hypothesis that certain types of collisions among fundamental particles occur at a rate that depends on the velocity with respect to absolute space.[2] Then, if we can measure the frequency of the collisions, we can break the tie and identify the winning version of Newtonian theory. (I'll ignore, for the moment, the disturbing

2. See Larry Laudan and Jarrett Leplin, "Empirical Equivalence and Underdetermination," *Journal of Philosophy*, 88, 1991, 449–472.

thought that the character of our measuring instruments is likely to leave us with an interval of values for the crucial parameter, rather than a single point.) It only takes a moment's thought to see that this is hopeless. For if the newly adopted hypothesis takes the rate of collision to be $f(v)$ where v is the velocity with respect to absolute space, and if the observed rate is $r = f(v^*)$, then, before we announce victory for the claim that the frame of the solar system moves with absolute velocity v^*, we should recognize that any rival proposal, for example the claim that the absolute velocity is w, can do exactly as well by rejecting the new hypothesis in the form given in favor of the rival version that takes the rate of collision to be $g(v)$, where g is chosen so that $g(w) = f(v^*)$. In other words, permanent underdetermination can be sustained because the allegedly tie-breaking hypothesis will have rivals that generate the results we want from any version of Newtonian theory we choose.

A better thought is to question the popular strategy of inferring from the fact that each of a set of rival theories yields exactly the same checkable consequences the conclusion that all will be equally well (or poorly) supported by any evidence we can acquire. It isn't immediately obvious that two theories making just the same predictions are equally confirmed by them. But the onus is now on the objector to set forth an account of empirical support on which the inference fails, and that is not easy.[3] Furthermore, even if the objector's point were correct as a general caution, it isn't apparent how it would apply to the case at hand. On what grounds can we distinguish Newtonian theories to argue plausibly that one is to be preferred to its rivals?

Of course, for post-Einsteinian physics, the debate about the underdetermination of Newtonian theory is moot (although I'll refrain from inquiring whether successor versions can be generated within contemporary treatments of space and time). Knowing that Newton's account gave way to relativistic physics, we can look back with equanimity on the earlier controversy, and even recommend an attitude that might have been taken. Before Einstein, we may suggest, there were two major possibilities: either Newton was right to insist on the existence of absolute space, and it was simply impossible to tell how the solar system moves with respect to absolute space, or his strategy was faulty from the beginning, and his laws of motion should have been viewed as picking out a privileged class of frames of reference (the inertial frames) of which the solar system is one, without any commitment to the existence of absolute space. It was always possible to adopt the concept of an inertial frame and to

3. If, for example, one takes degrees of empirical support to be probabilities and measures the degree to which e confirms h to be $\Pr(h/e)$, then, when both h and h^* logically entail e, an elementary application of Bayes' theorem shows that $\Pr(h/e) / \Pr(h^*/e) = \Pr(h) / \Pr(h^*)$. Hence the degrees of support in light of the evidence cannot be different unless there was a prior differentiation of the hypotheses. See John Earman, "Underdetermination, Realism, and Reason," *Midwest Studies in Philosophy*, 18, 1993, 19–38.

suppose that the solar system is among them, and prerelativistic physicists could have remained agnostic about the existence of absolute space and about the velocity of the solar system with respect to absolute space (assuming that it existed).

Turn now to the development of quantum mechanics. Schrödinger and Heisenberg articulated what they took to be distinct theories. On Schrödinger's account, states of a quantum-mechanical system are characterized by a function that evolves with time; the mathematical operators that correspond to observables (position, momentum, and so forth) remain constant over time. Heisenberg's version, by contrast, assigns each quantum-mechanical system a time-invariant state and supposes that the mathematical operators corresponding to the observables change with time. Theoretical physicists saw quite quickly that any experimental result derivable from one of the theories could be generated from the other, and that this arose from a systematic relationship between the mathematical structures on which Heisenberg and Schrödinger based their predictions. They concluded that these were not two theories but one, and that the underlying structure of the quantum world could be formulated in two different ways,[4] the Heisenberg "picture" and the Schrödinger "picture." This resolution is attractive not merely because it makes a problem about underdetermination go away, but because we have no theory-independent grasp of the notions Schrödinger and Heisenberg used in formulating their fundamental principles, so that it is extremely natural to take the systematic mathematical correspondences among these notions to reveal that the theories are intertranslatable, that they really say just the same thing.

Although these examples have sometimes motivated general theses in the philosophy of science, I want to draw two different morals from them. First, where cases of permanent underdetermination exist, there may be no systematic treatment to be given of them. Sometimes we may want to say what the physics community said about quantum mechanics: one theory, two different formalisms. On other occasions—among which may be the debate about motion with respect to absolute space—our appreciation of the fact that the language used has a meaning that isn't exhausted by the theory may incline us to suppose there is a genuine issue which we do not see how we could ever settle. Thus I resist any attempt to make underdetermination examples serve any grand 'ism'.

My second moral is that the belief in permanent underdetermination arises from a very special feature of these examples. Our justification for thinking that, however many experiments and observations we made, we would always find the theories in agreement rests on our ability to delimit very clearly the class of potential findings and to exploit the mathematical formulations of the theories

4. In fact, it can be done in infinitely many different ways.

to show that each can always be extended to yield the predictions generated by its rival. I now want to argue that this dependence on unusual characteristics of the influential paradigms should make us wary of the global underdetermination thesis, and that, in those rare instances in which we do face permanent underdetermination, the pragmatic attitude commended in my first moral indicates how we should understand the ideal of objectivity.

Geologists, chemists, biologists, and many physicists tend to be impatient when they hear about the problem of the underdetermination of theory by evidence. A common response is to declare that this is simply a philosopher's problem (in the pejorative sense), a conundrum that people with a certain quirky intelligence might play with, but something of no relevance to the sciences. That response is overblown, for, as we have seen, there are cases in which the problem does emerge in scientific practice (and others may well confront us if contemporary string theory fulfills its promise). Yet the sound instinct expressed in quick dismissal is a legitimate wish to be shown convincing examples across the range of scientific disciplines. To pick one case at random, what's the supposed rival to the hypothesis that the typical structure of DNA is a double helix with sugar-phosphate backbones and bases jutting inwards? If it's always possible to construct rival theories that generate exactly the same predictions and interventions, why are we so stumped when we try to think about alternatives that would account for the crystallographic data, the successful gene-splicing experiments, and all the rest of the vast array of consequences that have come from Watson and Crick's celebrated hypothesis?

Consider how hard it would be to defend permanent underdetermination in a case like this. We lack the general characterization of the relevant evidence that can be given in the Newtonian and quantum mechanical cases. Indeed, we know that attempts to specify the range of phenomena to which the double helix model could be applied were partial and myopic. Until the 1970s nobody would have guessed that we could develop techniques for making copies of DNA molecules with sequences to order, that we could insert these molecules into cells and watch those cells churn out selected proteins. Any guess we could now make about the limits of the evidence to discriminate rival accounts of DNA structure would be simply that—a guess. Moreover, even with respect to the available evidence, it's extremely hard to generate a serious alternative, precisely because there's no analogue to the free parameter in the versions of Newtonian theory. No mathematical transformations are available to generate a different hypothesis that will yield exactly the same predictions and interventions of the types we already know. In this instance, and in many like it, the underdetermination thesis seems to be based on philosophical faith.

Standard philosophical devices are, of course, available to generate "alternatives." With a nod to Hume, one might propose that DNA molecules have the standard Watson-Crick structure up to some chosen time in the future and, thereafter, have the bases pointing outward. Since the time can be selected in infinitely many ways, we obtain an infinite set of "rivals." All the "rivals" do just as well as the orthodox hypothesis with respect to the evidence we now have; those that fix the crucial time sufficiently far in the future may be permanently indistinguishable from the standard view with respect to evidence that members of our species could ever collect.

Alternatively, one could claim that permanent underdetermination is rampant by focusing on those theories that contain claims about the functional relations among quantities. Consider any such theory and any such claim. However many experiments we do, there will always be infinitely many (indeed uncountably many) rival functions that subsume the data we collect. Hence our predicament with respect to this theory will be one of permanent underdetermination.

Devices like these only expose the motivation for dismissing the global underdetermination thesis as a philosophical contrivance. There are, of course, interesting philosophical questions about whether one can discover satisfactory responses to Hume's famous worries about generalizing inferences or justifications of standard ways of fitting functions to quantitative data.[5] One way to approach these issues is to consider various strategies for pursuing inquiry and to argue that our standard ways of proceeding will succeed as well as any rival we can entertain. Another is to use mathematical learning theory to try to characterize types of problems that we may reliably solve as we sequentially obtain more evidence.[6] Any successes obtained in these enterprises would be welcome, but defense of the ideal of objectivity need not await them. For the point of that ideal isn't to announce solutions to the largest skeptical problems but to draw a distinction within scientific practice.

Underdetermination is cheap. You would have all the evidence you do for your current views even though the world had only existed for the past five minutes and had originated with the apparent traces of a lengthy history. (A similar hypothesis has occasionally been floated in attempts to square the evidence for evolution with a literal reading of Genesis.) When people appeal to the ideal of objectivity to distinguish the scientific apologists for Big Tobacco from the researchers who serve as their paradigms of proper detachment, they have no interest in whether or not the latter have convincing grounds for ruling out *every* rival hypothesis, including those that truncate the past. They tacitly suppose

5. In the latter case, for example, there is an interesting challenge to probabilistic notions of empirical support, since, when there are uncountably many rival hypotheses, it is impossible that all of them should be assigned non-zero prior probability. Does this make a certain sort of dogmatism unavoidable?

6. For pioneering work along these lines, see Kevin Kelly, *The Logic of Reliable Inquiry* (New York: Oxford University Press, 1996).

that there's a class of relevant alternatives and that fault lies in dismissing one of *these* without adequate evidence. Scientists who respond to philosophers' talk of underdetermination with a yawn (or worse) are denying that the mechanical devices we've briefly reviewed generate relevant alternatives. "Fake history" rivals, "chaotic future" rivals, "weird function" rivals inhabit the same territory as evil geniuses and brains-in-vats, and that territory belongs to philosophers. Whatever order philosophers can bring to their domain is welcome, but one can make the pertinent distinctions among scientists without awaiting their legislation—even without knowing how to draw the boundary that separates the genuine concerns (are we at rest with respect to absolute space?) from the don't cares.

Research scientists sometimes announce that a particular drug or procedure will aid in treating a disease and later receive criticism when it's discovered that they've ignored conflicting evidence or failed to run proper controls. Especially when their actions are designed to swell the coffers of the enterprise that pays them they are charged with a failure of objectivity. Contrast such behavior with the attitudes of the molecular biologists who accepted the Watson-Crick hypothesis. Critics of the ideal of objectivity now point out that, because of global underdetermination, there's a failure of objectivity here too. We ask for specification of the rival and a demonstration of permanent underdetermination. They reply that there must be one. We press for details. There is a long silence. Then we are offered a chaotic future in which molecular structures change after 2100 or one of the other mechanically generated alternatives. Is the failure to eliminate such "rivals" really tantamount to the oversights of the biomedical hucksters?

Although the global underdetermination thesis has been popular in efforts to impugn the ideal of objectivity, those who are sensitive to the history of the sciences often prefer a different line of argument. As we scrutinize the scientific debates of the past, we discover that they were not clashes between the clear-headed who perceived just where the evidence pointed and the prejudiced or muddle-headed who failed to do so. Rather the available evidence was thoroughly mixed, and each rival doctrine had its own successes and its own failures. Nor, so the story goes, did the situation change through the course of the controversy. At every stage there were successes and failures on both sides and choices were made by people who valued some solutions above others. Especially in the protracted episodes we call "scientific revolutions," people had to balance advantages here against problems there, and their overall decisions inevitably reflected a scheme of values. Changing one's mind could only be a "conversion experience."[7]

7. In talking of "conversion experience" I present one of the voices of Thomas Kuhn's influential *Structure of Scientific Revolutions* (Chicago: University of Chicago Press, 1962/1970), from which

A little history is good. We come to appreciate that Galileo, Lavoisier, and Darwin were not opposed only by ignoramuses, bigots, and fuzzy minds, but by intelligent defenders of alternative views who supplied challenging arguments. More history is better. For when we look more closely at the course of historical controversies, especially if we undertake the currently unfashionable work of analyzing the lines of reasoning, we discover that the disputes evolve, that positions are modified and new options emerge, that some problems are solved and others are generated. We find, in fact, that what was at one stage an impasse that made the tentative adoption of either of the rivals reasonable turns into a situation in which the balance of evidence is clear.

Yet even if it were agreed that scientific revolutions are resolved by collective leaps of faith, that would not deprive us of the ability to employ the ideal of objectivity more locally. Unless the critic can show that the features alleged to be present in the large debates permeate everyday scientific decisions it will be possible to use a tradition-relative conception of objectivity to separate some practices from others. (At this point the temptation for critics to invoke the global underdetermination thesis, or some analogue of it, is apparent.) Perhaps the force of the ideal would be diminished, however. For we would have to admit that scientists working within a tradition would be held to standards that were relaxed for the founders in whose footsteps they follow. The thought that each case of deviation from the ideal might be reconceived as an abortive attempt at a new revolution is disconcerting.

In any event, there is no need to examine if a tradition-relative ideal will serve provided that the hard cases, the major scientific revolutions, can themselves be seen as episodes in which decisions were reached on the basis of the evidence. An example will provide the general outlines of a story that can be told, so far as I can see, in all cases. It will also show us why the task of giving a formal account of evidence is so hard.

The debate between Lavoisier and the chemists who defended the phlogistonian theory of combustion occupied the better part of two decades. In 1770, most European chemists believed that combustion was a process in which the substance burned emits phlogiston to the air. By 1790, the overwhelming majority had adopted Lavoisier's new chemistry, which offered what has remained the standard view of combustion (that when things burn they absorb oxygen from the air). The dispute had a much wider focus. Both sides ventured hypotheses about the ways in which a variety of substances (acids, salts, gases) were composed of more elementary constituents. In the course of the dispute,

---

the argument of the present paragraph descends. It is important to recognize that the book contains a different voice, one that claims that scientists are ultimately persuaded by argument; Kuhn often rejected the attacks on objectivity launched in his name. Harmonizing the voices in his text is a difficult interpretative problem, and it is not my concern here.

the hypotheses of both sides were repeatedly modified, as new compounds were formed, previously unknown substances isolated. Intricate experiments were devised to pit the rival views about constitution against one another. Much of the discussion was complicated, as eventually became clear to all the participants, by the fact that the reactants being used often contained impurities. Hence there were rival accounts of why small amounts of some substances appeared in the products of certain reactions, and these generated suggestions about how these effects would disappear if the right procedures were followed in preparing the reactants.

Both sides wanted to offer a systematic account of the constitutions of a range of compounds in terms of the substances they took to be more elementary. In the early stages of the debate, each could point to a few successes, there were many instances in which one theory or the other was inapplicable or demonstrably unsatisfactory, and there were numerous avenues available to both parties to explore new variants. Lavoisier (and later his adherents) won the debate because he could articulate a systematic set of hypotheses about composition that generated predictions and interventions in an increasing collection of cases and because he could devise experiments that defeated the parallel efforts of the opposition. The complexities of this intellectual feat are quite extraordinary: following Lavoisier's path through the 1770s and early 1780s, one has to be constantly aware of the implications that modifications of his hypotheses have for a large number of predictions that he had previously made; at times, the number of phenomena to be considered is so large that it is hardly surprising that he was lost and had to retrace his steps; but what he achieved was an extensive series of arguments that eliminated all the plausible rivals to his own proposals, and, as these arguments came to be appreciated, more and more of his former opponents adopted his system.[8]

Presenting the logic of individual arguments is not difficult: given a particular set of background constraints, a specific experimental finding is shown to be inconsistent with the rival proposal or consistent with the favored proposal. Debate goes on, because there are many different ways of adjusting the background constraints, given the broader goal of achieving a systematic account of the composition of the pertinent substances. Those different possibilities have to be explored. Sometimes, however, an exhaustive search of the possibilities available to Lavoisier's opponents reveals no consistent option, and, lacking any further idea, they are forced to confess that this is an unsolved problem. As such unsolved problems mount for one side, and as previously recognized difficulties

8. This capsule summary does little justice to the details, which can be found in F. L. Holmes's classic study *Lavoisier and the Chemistry of Life* (Madison: University of Wisconsin Press, 1985). I have given a more extensive analysis in chapter 7 of *The Advancement of Science* (New York: Oxford University Press, 1993), where I also discuss in some detail the example of the Darwinian revolution. The point here is just to expose the structure of the resolution of revolutions.

for the other turn into successes, the scales begin to tip, and we appreciate how the evidence favors Lavoisier.

The account I have sketched makes it obvious why some scientific controversies should be protracted. The requisite explorations take time, and those who undertake them do not always find the shortest way through the labyrinths of possibilities. Furthermore, the fact that we find it hard to generate a precise account of scientific evidence that will apply to such complicated debates should no longer be surprising. Just as in the early stages of games of chess between grandmasters it may be difficult to tell which side is winning, so too with the great transformations in the history of science. Subtle accumulation of small advantages eventually tells, and, although the character of the individual gains can readily be understood, there is no clear and unambiguous way of tallying them. Numerous authors have written clearly and precisely about chess tactics—as there are scores of treatments of the logic of individual arguments—but the evaluation of strategic considerations is notoriously vague and qualitative. These differences are reflected in attempts to program computers that will compete with the very best chess players in the world, which, precisely because of the difficulty in coding strategic considerations, must rely on brute computational power as a substitute.

I conclude that neither the fact that major scientific controversies are protracted nor our inability to delineate a precise account of scientific evidence should undermine our confidence that the resolution of scientific debate on the basis of evidence is impossible. The ideal of objectivity need not be dismissed as a fond delusion. Hence there is no basis for believing that value judgments inevitably enter into our appraisal of which of a set of rival hypotheses (if any) is approximately correct.

So far, then, the discussion has favored one of the two images with which I began, the view of the scientific faithful. It is now time to articulate a different perspective.

FOUR

The World as We Make It

An old metaphor, bequeathed to us by Plato, conceives nature as an animal, at whose joints the objective inquirer should carve. Less flamboyantly, many thinkers, from the ancient world to the present, propose that nature has an order, a structure the sciences aim to expose. They maintain, for example, that chemical elements as we currently conceive them, that biological species, that diseases like diabetes and AIDS are real, and that there are no such things as the salts of medieval chemistry, the fixed species of pre-Darwinian biology, or such once-popular conditions as neurasthenia and hysteria. For thinkers in this tradition my accounts of realism and of the ideal of objectivity will appear too weak. The world comes to us prepackaged into units, and a proper account of the truth and objectivity of the sciences must incorporate the idea that we aim for, and sometimes achieve, descriptions that correspond to the natural divisions.

A concrete example will help bring these ideas into focus. Looking back into the history of medicine, we find it natural to say that there was no such thing as neurasthenia. To be sure, there were women who genuinely suffered and who were described by their doctors as "neurasthenic," but we now see them as a heterogeneous group, some with vitamin deficiencies, others the victims of social conditions that allowed them too little opportunity for activity, yet others with various afflictions we diagnose in very different ways. The point of denying the existence of neurasthenia is to insist on the heterogeneity: neurasthenia doesn't exist because neurasthenics turned out to have nothing in common with one another. Nothing? Well, we may say, nothing that distinguishes them from other women who would not properly have been classified as neurasthenic. Because they are so dissimilar from one another, neurasthenics do not form a "natural kind."

Some ways of classifying and the classes they pick out strike us as extremely artificial. Consider the class whose only members are the manuscript of *Finnegan's Wake*, Queen Victoria, and the number two. What do these things have in common? The response, "They share the property of being a manuscript of the last novel of an Irish expatriate writer who lived in Trieste and Zurich, or of being a female British monarch who reigned for more than sixty years, or of being an even prime number," invites the charge that this is no genuine property at all. There are, then, various ways to gloss the notion of an artificial classification. A classification is artificial when it groups together things that share no genuine property; when it is not a division in nature; when it lumps things together that are objectively dissimilar. Any of these interpretations will do to articulate the idea that, although it was once treated as a natural kind, neurasthenia turned out to be an artificial class.

We began with a metaphor, and it is not obvious how much progress we have made in arriving at a literal formulation of it. Invocations of "genuine properties," "divisions in nature," even of "objective similarity," are hardly pellucid. Because I distrust these notions, I am suspicious of the enterprises in which they are employed, and, particularly pertinent for the purposes of this essay, of the idea that there is an agenda set for our inquiries by nature. My aim in the next three chapters is to scrutinize that idea and to replace it with a very different view of inquiry, one that allows a place for human values and human interests in the constitution of the goals of the sciences. If we are to find our way between the unacceptable images of the first chapter, we need to resist the metaphysical encrustations routinely attached to the modest versions of realism and objectivity I have been defending. I shall start by opposing Plato's vivid image with a different picture.

Imagine a block of marble, one large enough to set a sculptor salivating. How many things are there here? One large block of marble, of course, but the question is notoriously ill posed. For there are many different lumps of marble inside the big block, many potential statues waiting to be released. Let's restrict our attention to statuesque representations of David. There are pieces of marble that take the shape of Bernini's version, of Michelangelo's, and of Donatello's (although his *David* was originally cast in bronze), and there are as many of each as we can distinguish different sizes. If Bernini (say) chips away at the block and produces his *David*, then he transforms the block into new objects (a statue and a collection of fragments and dust)—or, from another perspective, he changes the environment of a particular *David*-shaped lump that was already there, so that it now has air, rather than more marble, at its boundary. Were we to think of the marble as completely continuous, we could see it as containing an uncountable infinity of objects (many of which overlap).

The marble is a little piece of the world, and like the larger cosmos it can be conceived as divided up into objects in many different ways. Independently of our conceptions, those objects, those chunks of marble, exist. We draw (or chisel) the lines, but we don't bring the chunks into being. There is thus no determinate answer to the question, "How many things are there?" and no possibility of envisaging a complete inventory of nature. One highly unsatisfactory way of developing the point would be to declare that, prior to our conceptualizations, there are no objects. It is bad arithmetic to identify "very many" with "none," and it is equally bad metaphysics to suppose that the possibility of recognizing the boundaries of objects in different places is tantamount to supposing that there are no mind-independent objects at all.

There are many ways of developing concepts of things, events, and processes, and the different notions will pick out parts of nature that are (often) independent of us and our cognitive activity. There are uncountably many languages, some large number of which we could probably use, that would offer different concepts of objects, events, and processes. Virtually all of them would strike most of us as bizarre, and, it must be conceded, they would be quite inappropriate for our everyday purposes, for the formulations of the sciences, for persuading, entertaining, informing, and so forth. In principle, though, each of the languages could be employed to state large numbers of truths about nature. This is readily apparent from the case of the marble block. Imagine an alternative language—lumpspeak—which distinguishes precisely 53 lumps of assorted shapes within the block and assigns them numbers. Lumpspeakers can find convoluted ways of specifying in their own idiom the truths that ordinary speakers (of naturally occurring human languages) formulate about the block. Their true judgments about the spatial relationships among the various lumps—for example "Lump 23 abuts lump 41"—can be captured in natural languages by drawing distinctions we normally find quite pointless.

Some languages are richer than others, in the sense that one may have the power to express everything that the other can and more besides.[1] Whatever level we choose in the expressive hierarchy (or hierarchies), there will be a vast number of alternative languages with exactly that expressive power. The modest realism I have been advocating makes no claim that one of these languages is the uniquely right language at that level of expressive power for the purpose of describing nature. It simply recognizes that we can make sense of the notion of truth for natural and scientific languages. We can also make sense of the notion for any of the enormous array of alternative languages, and, since the truths of any of those languages can be formulated (with more or less difficulty) as truths in any of the others, there is no threat of conflict.

1. Obviously, comparative richness is not a complete ordering of all actual (let alone possible) languages, in that it is quite common for two languages each to have expressive resources the other lacks. Yet the quite banal extensions of language that often occur in the history of natural languages frequently produce later versions of those languages that are richer than their predecessors.

Those in the grip of Plato's metaphor will think we ought to go further. There is a *right* way to divide the world into objects, and it is encapsulated in the divisions found in the natural languages. Lumpspeak is eccentric, and some of its envisageable cousins, those that take part of Mars and my left thumb to constitute a single object, are downright mad. The insanity stems from the inability of these languages to reflect the boundaries of objects in nature. Human natural languages are, if not Nature's own, at least better approximations to it.

The line of reasoning just rehearsed plainly assumes all natural (and scientific?) languages share a common conception of the division of nature into objects, but I shall not quarrel on this score. For, while I agree that Lumpspeak and its wilder cousins tilt towards lunacy, I take that judgment to condemn them as poor choices for thinkers and agents with our biological and psychological capacities and with our interests. On the face of it, the thought that nature comes with little fenceposts announcing the boundaries of objects is an absurd fantasy, and the correlative concept of "nature's own language" is clearly a metaphor. If these notions are to be given any literal significance, the interpretation must reflect the view that some languages make it especially easy to describe our world or to reason about it. The latter formulations are in fact elliptical. 'Ease of description' can only be understood in relation to beings with particular capacities and particular aims. Making an entire inventory of the world is, I have claimed, not an option for us—and neither is offering a complete description. Insofar as we attempt to describe, to predict, to explain nature we are inevitably selective. Hence, in providing a literal interpretation of "nature's own language," we are driven to the conclusion that the languages we use are apt for the description of nature in the sense that they are good for *creatures like us* to formulate the kinds of descriptions of the world *that we care about.* In other words, human languages strike us as distinguished from their weird rivals because of the banal fact that these languages are good ways for us to achieve our purposes.

This is by no means the end of the argument. Staunch champions of the objective correctness of particular languages will surely propose that scientific discourse attempts to achieve purposes that are not contingent products of human history but constitutive of the goals of inquiry. That is, they will view our selection of some aims as an expression of our recognition of what it is to attain knowledge. Their attempts to defuse the relativity to human concerns will occupy us later (most explicitly in chapter 6). I want, however, to close this initial skirmish with a different perspective on the possibility of variant conceptions of the constituents of nature.

One natural way of responding to concepts of object that we find bizarre is to appeal to spatial, or temporal, connectedness. To count some part of Mars and my left thumb as parts of a single object is mad, we might think, because these "parts" are separated by an expanse of space that is taken not to belong to the object. Similarly, if we imagine a single organism whose life consists of the last decade of Bertrand Russell followed by the first decade of the family dog, we may express the judgment that this is not a natural object on the grounds that there is a temporal gap between Russell's end and the dog's beginning. At first glance, the everyday objects that serve as paradigms do seem to satisfy requirements of spatial and temporal connectedness. Yet even such mundane phenomena as broken chairs remind us that the parts of a single object do not have to be spatially connected at all times. Reflecting a little further on such entities as the United States and the United Kingdom, it becomes evident that spatial connection is not necessary. Of course, these examples are different in other respects from the "object" that comprises part of Mars and my thumb, for there is an obvious point to treating the separated part of the chair as part of the whole, to taking Alaska and Hawaii as parts of the United States, Northern Ireland, Guernsey, Skye, and the Isle of Wight as parts of the United Kingdom. The chair as a whole has practical and aesthetic functions; the countries are each subject to a common government. Yet to appeal to such considerations is to refer, once again, to our interests and purposes.

Even ordinary objects, with their apparently unproblematic spatial boundaries, dissolve under the lens of microphysics. Not only is it far from obvious that traditional conceptions of objects are appropriate for characterizing the entities physicists currently hypothesize as the basic constituents of matter, but, even granting their applicability, the issues of characterizing the spatial boundary of an object and the type of spatial connection that applies within it are far from trivial. Nor does contemporary physics lend much comfort to the view that objects should conform to a temporal connectedness requirement. Conceiving the world as a four-dimensional space-time manifold, the distinction between objects and events is blurred. Yet we readily recognize temporally disconnected events: as I write, more than three thousand fortunate New Yorkers have just attended a production of Wagner's *Ring*. Further, unless time-travel is a conceptual impossibility, we ought to countenance objects whose world-lines are broken, who live for a while, vanish, and emerge later just as they were at the time of their disappearance. I have no doubt spatiotemporal connectedness is a convenient condition for organisms like us to use in dividing nature into objects, but only a little imagination is required to appreciate that, had we been differently constituted, we might not have imposed this requirement widely (or even at all) and that, constituted as we are but with different interests, we might have given greater scope to divisions of reality that flout it.

Once again, I want to caution against overinterpretation. Different ways of dividing nature into objects will yield different representations of reality. Users of different schemes of representation may find it difficult to coordinate their languages. Properly understood, however, the truths they enunciate are completely consistent. What is true in Lumpspeak is not at odds with what is true in English, although Lumpspeakers may appear to us to employ a perverse and unmotivated way of formulating the "important" truths about nature. That, of course, is because they have a rival conception of what counts as important.

Those in the grip of Plato's metaphor suppose not only that there is a privileged way to divide nature into objects but that there are natural ways to assort those objects into kinds. They recognize that, at different times and in different places, people have used alternative classifications. Three centuries ago, even educated people counted whales as fish and, famously, some contemporary people only count as birds those feathered creatures that actually fly. Faced with this variety, ambitious realists argue that the sciences teach us how to abandon faulty taxonomies and to recognize the "real similarities" in nature. Here too I see a dependence on human capacities and human purposes.

Consider our divisions of organisms into species. Bitter intellectual battles have been waged around the question of the correct principles on which biological taxonomy should be based. The most influential doctrine remains the "biological species concept," so called because it appeals to discriminations made by organisms themselves: species are clusters of populations that consist of freely interbreeding organisms and that are reproductively isolated from populations in other such clusters.[2] Unfortunately, it turns out to be rather a delicate matter to characterize a population without already deploying some species concept and to say just how much interbreeding is compatible with reproductive isolation. Yet for many biological purposes the biological conception is well enough understood. For many—but not for all. Botanists who deal with plants in which reproduction is asexual or in which hybridization is frequent usually appeal to morphological characteristics of their specimens, and it would be an odd interpretation of their practices to view them as aiming for a division that focused on traits that are important in classifying some related group in which reproduction is sexual; typically, they have more immediate concerns. For some ecological projects, the ecological role of the organisms can override criteria of reproductive community or reproductive isolation. Paleontologists, attempting to bring order to a fossil lineage on the basis of an imperfect record that only retains the hard parts of organisms, necessarily appeal to morphological criteria. Finally, in the study of microorganisms, in which precise molecular specifications are possible, viruses and bacteria are often grouped on structural principles. Contemporary biology seems committed to pluralism, as different investigators use the classifications best suited to their needs.[3]

2. This conception was pioneered by Ernst Mayr in his *Systematics and the Origin of Species* (New York: Columbia University Press, 1942) and in a large number of subsequent books and essays. The influence of the conception stems from Mayr's ability to show how his notion is useful for understanding many aspects of those organisms in which human beings are most interested.

3. For more extended discussion, see my essays "Species," *Philosophy of Science*, 51, 308–333, and "Some Puzzles about Species," in *What the Philosophy of Biology Is*, ed. Michael Ruse (Dordrecht: Reidel, 1989), 183–208.

Contrast two different biomedical endeavors. A great advance in the study of malaria resulted from the ability to differentiate between two groups of mosquitoes that closely resemble one another, so-called sibling species. Before researchers understood that they were confronting two interbreeding populations, reproductively isolated from one another, they found it hard to understand why malaria was prevalent in some regions and absent in others, all of which were infested with indistinguishable mosquitoes. In this instance, morphological criteria are worthless, and the biological species concept enables us to distinguish the harmless mosquitoes from those whose bites are more deadly. Yet, in the study of Lyme disease or of AIDS, the situation is very different. Here the bacterium in the one instance and the virus in the other are identified through precise molecular criteria, as researchers concentrate on discovering the structures that enable the disease vectors to attack human bodies. It would be absurd pedantry to insist that a single way of classifying organisms must take precedence and that one of the taxonomic schemes is "unnatural." For the purposes of the classification are both obvious and well motivated: researchers want to divide up the organisms in ways that help combat human disease. Once again, the partitioning of nature accords with our interests and, in a less obvious way, with our capacities.

Even within biology, more general divisions of organisms are controversial. Sometimes it is suggested that classifications of organisms into genera and the higher taxonomic categories ought to conform to an evolutionary principle: organisms that share a more recent common ancestor are more closely related than organisms whose most recent common ancestor lies further in the past. Nobody should question either the value of that principle in evolutionary studies or the interest of such studies. Nevertheless, ecology, structural biology, and other subdisciplines have their own concerns, and it would be folly to insist that their classifications ought always to be imposed by reference to a history that may be both irrelevant and conjectural.

As we move beyond the contexts of biological research, the folly becomes even more obvious. Amateur gardeners and even horticulturalists have an uneasy relationship with the distinctions of professional taxonomists. They divide the organic world into trees and shrubs, flowers and weeds; they pick out cedars among the evergreens, and they demarcate chollas from prickly pears: in none of these instances do their classifications correspond neatly to the biological taxa.[4] Leaving the garden for the kitchen, the story is the same. Neither cooks nor those who grow their ingredients care much for the fact that the garlics and onions belong to the family that includes the lilies.

We divide things into kinds to suit our purposes. Yet, one might object, successful classification ought to be informed by the divisions drawn within the sci-

4. See John Dupré, *The Disorder of Things* (Cambridge, Mass.: Harvard University Press, 1993), chap. 1. Many of the points in this paragraph draw on Dupré's excellent discussion.

ences for the fulfilment of scientific purposes. The brief discussion of species concepts has already begun to chip away at this objection, but a more sustained response to it will occupy us later (chapter 6). A different attempt to defend the privileged status of some classes is to suggest that the messiness of biology should not blind us to the clean lines of division marked out in more fundamental sciences. Physics and chemistry furnish examples of genuine natural kinds.

Twentieth-century physical science taught us to divide the elements by atomic number (gold is the element with atomic number 79), to organize compounds according to their three-dimensional composition from elements, and to understand general chemical notions in terms of the basic properties of microphysics (acids are electron pair acceptors). These classifications are felt to have a special status because the properties deployed in marking out the divisions are those that play a systematic explanatory role. More exactly, the superficial similarities and differences on which our ancestors fastened in their primitive efforts to catalogue substances are explained in a unified fashion by showing how they flow from the properties that underlie current scientific divisions. The yellowness, shininess, and ductility of gold is derivable from its being the element with atomic number 79, and, similarly, the atomic numbers of the other elements ground their manifest properties.

In the history of the physical sciences the search for the "inner constitutions" on which observable characteristics depend has been important and challenging. From a contemporary perspective, it is easy to see that search as exposing the "fundamental characteristics of things." Yet the character of the process bears scrutiny. Certain microphysical structures are seen as grounding an important similarity among things because those structures are a common constituent in the causal processes that generate a number of superficial features: the state of being an electron acceptor plays a causal role in reactions between the things we call acids and a host of other substances. Notice, however, that the idea that these microstructures play a systematic unifying role depends on the *prior* identification of a class of manifest properties. Perhaps because of our sensory and cognitive capacities, perhaps because of interests that people have developed, either naturally or as a result of the accidents of human history, we focus on certain aspects of the world that intrigue us. Relative to our concern with these features, the microphysical structures that underlie our physical and chemical classifications do indeed have a privileged status, for the features in question turn out to be common constituents of causal processes that generate the fascinating surface properties. Beings with different constitutions or with similar constitutions but different interests might begin with an alternative collection of manifest features, and their inquiries might single out alternative underlying microphysical structures, causally implicated in the cluster of characteristics that concerned them. Or perhaps the common causal factors would not be *micro*physical at all. Even in the case of the physical sciences, the alleged objective similarities carry a tacit relativization to our capacities and our interests.

I suggest that only a failure of imagination stands in the way of an appreciation of this point, but it must be admitted that I have not articulated a detailed fantasy to show what the rival set of categories would be like. In consequence, the defender of privileged divisions in nature can insist that *any* collection of initial surface characteristics would eventually lead to the same underlying classification. But why should that be so? In the biological sciences, actual practice reveals examples of how different purposes demand different categories. To insist that alternative physicochemical interests, bizarre as they might be by our own lights, would inevitably require just the same division on microstructural lines looks like an article of faith.

So, do we construct the world? In the sense often intended in fashionable discussions, we do not. There is all the difference between organizing nature in thought and speech, and making reality: as I suggested earlier, we should not confuse the possibility of constructing representations with that of constructing the world. If the claim that the stars are a human construction is to mean anything remotely plausible, it can only be that there are alternative ways of dividing nature, that some of them would not recognize stars as objects, or would classify astronomical objects without using the category *star*. On that interpretation, to repeat, the stars are independent of us—and so too are the entities of an indefinite number of other schemes of classification.[5] Like the statues in the marble, they are all already there, even before we draw their boundaries.

Yet we do construct the world, or, at least, quite a lot of it. The stars are too remote to suffer the consequences of our activities, but, closer to home, matters are different. As I look out of my window, my gaze meets a vast number of man-made objects. Even the "natural" things, the trees lining the avenue, for example, are as they are partially as the result of human intervention; not only are the individual trees pruned and shaped, but the seed from which they are grown was obtained after generations of horticultural selection. The effects of human activity would be less obvious were I in a more bucolic setting. Less obvious, but not absent. The pastoral landscape and the prairie take their present forms because an ecosystem has been maintained or modified through processes in which human beings are intimately involved. Human hunting, clearing, planting, and pollution shape almost every locale on every continent.

On the face of it, these two senses of "constructing the world" have little to do with one another, and to treat them together is to perpetuate a silly ambigu-

5. In a penetrating article, "Putnam's Pragmatic Realism," *Journal of Philosophy*, 90, 1993, 605–626, Ernest Sosa raises a problem with what he calls the "explosion of reality." My approach here accepts the explosion, but tries to defuse the problem.

ity, a bad pun. Matters are not so simple, however. For our classifications, the ways in which we draw boundaries around pieces of nature to suit our capacities and interests sometimes lead us to modify the world as we find it. Nations have gone to war because they view pieces of land adjacent to their coastlines as intrinsically part of their territory, or because they have seen speakers of a common language as fellow nationals.

Imagine a different possible history for our planet. The counterparts of *Homo sapiens*, the most intelligent organisms evolution yields, have rather different interests from our own. Perhaps they even have different sensory and cognitive capacities. Their history of culture and intervention proceeds for just the same time as ours. At the later stages of this process, will it be true that our counterparts merely differ from us in the classifications they bring to the same nature? I think not. Nature itself will have been modified by their interactions. They will live in a different world in the perfectly comprehensible sense in which we think of contemporary industrialized societies and our pastoral ancestors as living in different worlds.

The impact of categories on reality by way of human intervention is more evident in the biological sciences than in the physical sciences and most striking in those areas of inquiry in which we study ourselves. Developing a classificatory scheme in which individuals are seen as insane, or as homosexual, or as belonging to different races has effects so ramified as to make it difficult for later generations to undo the consequences of earlier decisions. Not only are such classifications embedded in a network of institutions, laws, and artifacts, but they also foster self-conceptions that excite or inhibit actions. The resultant pattern of human activity can then easily reinforce the judgment that the initial classification marks an important natural division.

Increasing understanding of the genetic variation within and between populations of our species has made it abundantly clear that, from a biological point of view, there are no important differences among the groups traditionally marked out as races. The genetic distinctions are trivial, a matter of a small number of differences in the frequencies of some alleles. Variation within races, even on the traits commonly taken as indices of racial difference, is considerable. Yet, even if all human beings were to recognize, today, that the notion of race was bad biology and that it should be abandoned, that would only be a start on the project of undoing the effects of our past practice of dividing human beings into races. The history of actions based on acceptance of racial difference, most prominently but by no means solely in the United States, has left its mark on the relative conditions of different people, in socioeconomic terms, in the availability of opportunities, in the character of the neighborhoods in which people live and the ways in which they lead their lives, and, not least, in the conceptions people have of themselves, of their families, and of their past. In a very real sense the world we inhabit is as it is because of classifications adopted by our predecessors, including concepts we would like now to repudiate.

Categories are consequential. Accordingly there is important work—philosophical? archeological? historical?—to be done in reconstructing the ways in which our most influential divisions were constructed and how they have left their mark on the world we inherit. Through such excavations, we can hope to improve our classifications and to make changes that will help us realize the goals we currently set for ourselves. If in the past our classificatory practice has had effects that are now confining, for some subgroup or for all of us, sufficient understanding of that practice may enable us to liberate ourselves. What we have constructed, we may be able to dismantle.[6]

The world we make is not formed by some mysterious process of metaphysical construction. In mundane causal ways, human beings modify their natural and cultural environments. Our ways of dividing up nature play a causal role in the process. So the world is partly as we make it, but that thoroughly unmystifying fact should not call into question any of the modest realist points of earlier chapters. An analogy will help focus the view I am advocating and will prepare for further development of it.

6. My archeological metaphor in this paragraph deliberately resonates with the ideas of Michel Foucault. As I interpret him, Foucault's main aim was not to propose some general relativism in metaphysics and epistemology, but to identify the ways in which past choices have framed contemporary categories and projects. For a similar approach to Foucault, see Gary Gutting's introduction to *The Cambridge Companion to Foucault* (Cambridge: Cambridge University Press, 1994), 1–27.

FIVE

Mapping Reality

SOME SCIENTISTS DO NOT SPEND THEIR LIVES formulating equations or even inscribing sentences. The products of their labor are pictures, or diagrams—or maps. In both geology and genetics, map-making is a respected activity, but I want to clarify the picture of the sciences I have been developing by looking at the core field, the academically rather unfashionable discipline of cartography.[1]

The history of map-making illustrates the modest realism with which I began. Consider some of the maps of our planet offered by the geographers of the past, maps of the entire globe. Later maps appear superior to earlier ones in two major respects. First, they include entities that were previously omitted, the New World and Australasia being the most striking examples. Second, their depictions of the spatial relations among the entities commonly represented are more accurate; the margins of the various countries follow actual coastlines more closely. We make these judgments without believing that *any* of the maps ever produced is completely accurate, even while admitting the possibility that earlier maps might occasionally deliver a more accurate representation of some local features, and that the kind of convergence we appreciate visually need not be monotonic.

Past explorers might have had reason to think their maps were accurate because they were able to use them to navigate successfully (although there were, of course, any number of failures). The maps that superseded their charts have

1. Philosophers have sometimes turned to maps for comparisons with science. One particularly fruitful exploration, unjustly neglected in subsequent discussions, is Stephen Toulmin's treatment of the topic in his *Philosophy of Science: An Introduction* (New York: Harper and Row, 1953).

enabled their descendants to achieve success more systematically across a far wider range of voyages. In retrospect, we can also see why the earlier expeditions prospered to the extent that they did. Some features of the older maps are preserved in their later counterparts because, even though the old cartographers were not entirely right (and sometimes, of course, they were wildly wrong), they managed to achieve approximate accuracy about particular aspects of the globe. When they planned journeys that depended just on these features, they were able more frequently to reach their destinations.

Cartography displays in a particular instance just the type of progress and just the relation between success and accuracy that the modest realism of chapter 2 claimed for the sciences generally. Yet, when we think more carefully about the variety of maps—even the variety of maps of the earth's surface—we recognize complications. Map-makers are invariably selective. They introduce conventional elements, and, in consequence, standards of accuracy vary. The most obvious convention in the case of global maps concerns the way in which the three-dimensional relations are projected on a two-dimensional surface. It would be foolish to protest that a map of the globe using the Mercator projection is inaccurate because it makes the area of Greenland appear much greater than it is (relative, say, to South America)—just as foolish as denouncing a map for its uniformly pink coloration of the British Commonwealth. Associated with any map there are conventions that determine which aspects of the visual image are to be taken seriously.

Expanding our purview to embrace a variety of maps, we understand how maps designed for different purposes pick out different entities within a region or depict those entities rather differently. One map of a California resort region may display major roads, sports facilities, restaurants and services surrounded by a sparsely marked expanse of green, grey, and blue; another, designed for the serious backpacker, may show the roads only as conduits to the wilderness, while lavishing detail on the courses of streams, the sinuosity of the tree-line, the contour lines, and the trails. The shape of the same stretch of coast may be drawn differently—because of decisions about the coastal margin—in maps intended for the casual yachtsman, the holiday-maker in search of secluded bathing spots, the marine biologist, the geologist concerned with the fault structure, and the urban planner. What counts as an omission or an inaccurate spatial representation depends on the conventions associated with the kinds of maps, and, in their turn, those conventions are in place because of the needs of the potential users.

It might appear there is a limit to the variety of maps, some minimal set of conventions common to all, and some common standard of accuracy stemming from these conventions that every map is required to meet. So, for example, one might propose that if a map represents as collinear geographical features that do not lie on a line, then the map is ipso facto inaccurate. The proposal is mistaken.

Consider the map that figured in the scenarios of the second chapter. If practical success in navigating is to serve as our test of accuracy, then the map of the London Underground must count as accurate—for it figures in the successful activity of tens of thousands of people each day. But the map lies, in portraying as collinear places that do not fall on a straight line through space: the stations on the Central Line share a common Line but no common line.

These elementary reflections can help us to a more precise account of the conventionality and the accuracy of maps. Think of a map as a visual display coupled to a set of conventions. The set of conventions divides into two parts, the *intended content* and the *reading conventions*. The intended content of the map consists of the region and the types of entities and properties that the map intends to portray. The reading conventions link items in the visual display to those entities and also specify which features of the display do not correspond to any aspect of nature. In some instances, they will divide lines in the visual display into meaningful units and specify how these units correspond to parts of nature. Other conventions stipulate which spatial relations in the visual display correlate with spatial relations in the world. Yet others say that an aspect of the visual display is to be ignored. (Thus, old-fashioned maps of the globe, on which the British Commonwealth is colored pink, do not carry the information that Canada is uniformly pink.)

Consider again the map of the London Underground. The domain of the map (the region identified in the intended content) is London, and the objects of interest include the various Tube stations and the railway lines connecting them. The intended content identifies properties that are of interest, such as the relational property of two different lines being connected by a walkway (more colloquially, the map picks out stations where people can make connections). The spatial relations of special concern include being connected by the same railway line and being adjacent along the same rail line—but not being collinear in physical space nor any metrical relations among points depicted.

The reading conventions for this map connect dots and lines with entities in London: the dot marked "Clapham Common" is linked to a station in South London, the horizontal red line stands for the Central line, and so forth. Those conventions also tell the competent reader which parts of the display shouldn't be taken seriously, not just the coloration but also the ordinary conventions about points of the compass (the map doesn't inform us that Notting Hill Gate is exactly west of Oxford Circus).

I belabor the relatively obvious in order to show how to make sense of the notion of accuracy. Suppose that, in the visual display, two elements, identified by the reading conventions as bearers of meaning, stand in a particular spatial relation. According to the conventions, there are items in the domain of the map that correspond to those elements and there is a real-world counterpart of the spatial relation. The map is accurate, in this respect, if the two items from the domain stand in the counterpart of the displayed relation. For the London Underground, part of the map's accuracy consists in its depiction of Victoria and Green Park as connected, without intermediates, by a blue line, and, in light of the reading conventions, this signifies that Victoria and Green Park are adjacent stations on the Victoria Line—as indeed they are. The map is also accurate in showing a large black circle at Notting Hill Gate, signifying that it is possible to change there between the Central Line and the Circle Line. So it goes. On the account I have offered, the map of the Underground is not *approximately* accurate. It is exact.

This may be unusual in that this particular map is thoroughly discrete and finite. More typical maps have reading conventions identifying a large infinite number of signifying elements. Consider, for example, an outline of Manhattan island, accompanied just by a pointer indicating North. Any connected segment of the line can be regarded as a meaningful element, and, given the reading conventions, we can appraise the accuracy of that element in terms of the conformity between the shape and orientation of the line and the shape and orientation of the pertinent portion of the island's margin (which is, of course, itself conventionally fixed). In this way, the map is equivalent to a truly enormous number of claims about spatial relations (continuum many): a picture is not worth a thousand words, but rather a staggering infinity of sentences. Further, although the map says many things that are incorrect, it also expresses an infinite number of true statements, for there are infinitely many truths of the form "A is within φ of being θ from due North of B," where A, B are places on the Manhattan shoreline and θ, φ are angular measurements. Like most scientific theories, the map, taken as a whole, is false, even though it contains large amounts of truth.

There is no problem, then, in talking of the accuracy of maps and hailing some maps as accurate in some (even all) respects. Our realist commitments are, however, perfectly compatible with recognizing the fact that human interests change and, in consequence, maps are drawn with very different reading conventions. We now have little use for the Tudor maps that showed sheepherding trails and the boundaries of manors. Nor do we expect that our maps of the globe will display the significance of Christ's passion. The reading conventions for many older maps are very different from those of the present, but the change should not surprise us. Reading conventions identify the ways of dividing the spatial domain that are of interest to the map-maker, and those conventions depend on the goals and the institutions of the society in which the map is to be used.

There is no unique correct way for a map of the globe, or of some smaller region, to draw boundaries. Sometimes part of the purpose is to recognize political divisions, to demarcate nations, states, counties, administrative districts from one another. Sometimes we are interested in streets, roads, and highways. On other occasions the significant boundaries show the movements of migratory animals, the zones of common climate, the distribution of plant species, the topography of a mountain range, the extent of public and private lands, the successive inundations of a flood, the distribution of areas with specified population densities, incidence of disease, or availability of minerals. Which features are crucial as landmarks in drawing these alternative divisions varies from case to case: the botanist studying the distribution of arctic flora may have no use for any background markers at all, the hiker's needs may be met by a topographical map that indicates no more than access roads.

I argued in the last chapter for the analogous thesis about the sciences generally. Our ways of dividing up the world into things and kinds of things depend on our capacities and interests. The history of map-making extends the point by showing how cartographical conventions and divisions evolve in response to changing human purposes. Maps lose their place in human lives as the projects they once served are superseded, and those maps then retain value only for the historian. The intricate relations among Tudor manors and the minutiae of eighteenth-century waterways have no purchase on contemporary actions, and we can easily envisage that our descendants may find maps that reflect our administrative distinctions—and even the much-used map of the Underground—equally beside the point.

The map-makers' task is to produce maps that are pertinent to the enterprises and interests of their societies. By the same token, I suggest, the aim of the sciences is to address the issues that are significant for people at a particular stage in the evolution of human culture. Languages are fashioned to draw those distinctions that are most helpful in carrying out the lines of investigation those people want to pursue. As the history of cartography reveals a succession of maps with very different reading conventions, so too the history of the sciences generally should disclose a succession of languages framed, often imperfectly, to the pursuit of inquiries that appear, at the time, most important.

At this point we confront again an issue initially joined in the last chapter. Surely there is a grand scientific project constitutive of inquiry at any time, in any place, independent of culture, social institutions, or mutable human concerns? Whatever language, or compendium of languages, is apt for this large purpose will mark out privileged divisions in nature. It will identify the *real* natural kinds, the *genuine* objects and properties. Scientific inquiry aims to learn this language and to enunciate in it the basic truths about nature.

Although this conception of an overarching aim for inquiry has been influential in most discussions of the sciences, I am skeptical.[2] My skepticism surfaced in the last chapter, and it will be articulated further in the next. For the

2. I was not always so. See *The Advancement of Science*, especially chap. 4.

moment, however, I want to motivate it by drawing on the special case of maps.

Imagine a philosopher of cartography devoted to the idea that map-making has context-independent goals. A natural way to present the philosopher's thesis would be to suppose that the goal of cartography is the construction of an *ideal atlas.* The maps actually produced in human history are a selection of sheets from this atlas (to the extent, that is, that they are accurate). Of course the selections actually made are informed by interests that vary from group to group, epoch to epoch, but behind the contingent choices stands the inclusive ideal. Individual maps are significant because they belong to a hypothetical compendium, towards which we aim but which we shall never achieve.

What exactly would this ideal atlas be? It's implausible to suppose that it contains a single map that reveals all the spatial relations on our planet, for, as Lewis Carroll pointed out long ago (in *Sylvie and Bruno*), nothing could substitute for that except the terrain itself. Rather, the view must be that there are certain fundamental kinds of maps from which all spatial information can be generated, and that they collectively provide a unified presentation of the wide diversity of kinds of knowledge drawn from our actual ventures in cartography (and, presumably, projects we might have undertaken). But now a new question arises. What kinds of maps furnish information to advance any conceivable human project? Simply surveying the vast diversity of maps produced in actual human history—from which I have drawn a tiny sample—exposes the difficulty of reducing the atlas to any manageable compendium. Moreover, for any atlas we can envisage, we can easily conceive of projects that would require maps of different types—for example the project of recognizing the distribution of copies of the atlas itself.

There is no good reason to believe in the ideal atlas. A much more straightforward approach to the variety of maps, prefigured in my earlier expression of skepticism, is to relativize the notion of cartographical significance to communities, seeing some kinds of decisions about what to represent and how to represent it as the results of central aspects of those communities' ways of life. We would abandon the idea that cartography is governed by a context-independent goal. Perhaps we should lose similar baggage in thinking about the sciences generally.

The analogy between cartography and science invites a further step. Current ventures in map-making often carry the traces of past endeavors. To understand why present maps take the forms they do, we need to recognize the ways in which past projects of map-making have led to modifications of the part of nature that is mapped. Cartography generates a counterpart of the thesis that classifications may play a causal role in the reshaping of reality.

Consider a straightforward example in which map-making has contributed to the alteration of the physical environment and the development of various pieces of technology. Backpacking Californians use topographical maps to explore the wilderness of the High Sierras. Older maps (together with guidebooks, lightweight camping equipment, and so forth) made it possible for more people to experience the beauty and solitude of the mountains. In consequence, certain lakeshores became degraded from over-camping, the foraging habits of animals (most obviously bears) were disturbed, and hikers seeking solitude were pressed to explore higher altitudes. Forest rangers have responded by marking more clearly (and sometimes widening) trails once viewed as "cross-country routes" for mountain climbers. At the same time, as backpackers ascend above the tree line, pursued by bears who are ever more adept at liberating their food, new kinds of backpacking technology have been introduced, such as the "bear boxes" installed at some wilderness sites, or the plastic canisters that hikers can carry to protect food. The maps of today show more detail for the more remote elevations than did their ancestors, as well as recording the changes caused by human activity, and we can envisage that the maps of the future may display information about food storage that would have been irrelevant a generation ago. This simple example shows quite clearly how the full story of why one set of conventions is chosen must include the past choices of map-makers and the projects their maps made possible, for those maps and projects influence the desires of later map-users, the resources available to them, and even the character of the terrain that they will explore.

So it is, I suggest, with the sciences generally. Like maps, scientific theories and hypotheses must be true or accurate (or, at least, approximately true or roughly accurate) to be good. But there is more to goodness in both instances. Beyond the necessary condition is a requirement of significance that cannot be understood in terms of some projected ideal—completed science, a Theory of Everything, or an ideal atlas. Recognizing that the ideal atlas is a myth, I hope to have provoked concerns about the analogue for inquiry generally. A rival vision proposes that what counts as significant science must be understood in the context of a particular group with particular practical interests and with a particular history. It further suggests that just as maps can play a causal role in reshaping the terrain that later cartographers will depict, so too the world to which scientists of one epoch respond may be partially produced by the scientific endeavors of the past—not in any strange metaphysical sense but in the most mundane ways.

I've offered a motivational analogy, not an argument. Fans of the traditional idea that there is a context-independent aim of inquiry could accept everything I have said about maps, their accuracy, their conventionality, and the sources of their significance and also argue that map-making is intimately bound up with practical projects, so that there are no implications for the significance of pieces of theoretical science. To claim that maps are invariably drawn for specific practical ends would be overstatement—historians often employ maps to advance our understanding of the past—but I shall not try to move beyond my motivational exercise to extend the argument in this direction. The analogy helps to frame the main issue that will occupy us: If science is indeed different, what is the genuine counterpart of the admittedly fictitious ideal atlas? Can we provide an account of the goals of inquiry, a specification of what constitutes significant science, that will apply across all historical contexts, independently of the evolving interests of human beings? Let us see.

SIX

Scientific Significance

THE SCIENCE STORY OF 1997 centered on a Scottish sheep. In an article widely discussed in scientific journals and in the broader press, Ian Wilmut, a researcher at an agricultural station near Edinburgh, reported the birth of Dolly as a result of nuclear transplantation. Wilmut had extracted the nucleus from an egg taken from one ewe, replaced it with the nucleus from an udder cell of another ewe, implanted the resultant egg, and allowed gestation to proceed. His report was accepted as correct at the time, and, despite challenges, it has been upheld since.[1] But why all the fuss? Why is it a significant piece of science to show that a mammal can be born in this way?

An obvious part of the answer recognizes the practical importance of Wilmut's work. This doesn't lie in the absurd fantasies about cloning people that have been widely touted in newspapers and popular magazines. Consonant with the character of his position, the primary practical import of Wilmut's achievement lies in its opening up the possibility of breeding domestic animals with desired characteristics (resistance to common diseases, preferred musculature, and so forth). At the junction of animal husbandry and medicine, researchers also envisage the possibility of modifying the genome of a future nuclear donor, inserting alleles to direct the production of a useful drug, and obtaining a flock of clones whose milk would be laced with this substance. Farther from such everyday benefits is the potential use of nuclear transplantation to generate genetically identical mammals (probably mice) with some particu-

1. The challenges focused on the possibility that Dolly did not share the nuclear genetic material from the adult female Wilmut assigned as her "nuclear mother." Any such doubts have now been resolved by DNA sequencing. See D. Ashworth et al., "DNA Microsatellite Analysis of Dolly," *Nature*, 394, 1998, 329.

larly interesting genotype, one associated, say, with the analogue of a recalcitrant human disease, so that the effects of this genotype can be systematically explored.

This last implication bestows on his work an indirect import for inquiries in theoretical biology. But there's a more straightforward connection, one noted in the forthright appraisal of the commentary that accompanied Wilmut's original article: "The results are of profound significance."[2] The judgment is supported in the commentary's opening sentences: "A hoary old question that has interested developmental biologists for years has been the continuity of the genome during animal development. Put another way—do growth, differentiation and development of the embryo involve irreversible modifications to the genome in somatic cells?" Since the 1950s, developmental biologists have understood that virtually all the cells of an organism contain the same complement of nuclear genes. At the same time, it's clear that cells differentiate into types with very different properties—muscle cells, blood cells of various types, liver cells, neurons, and so forth. Given our understanding of interactions between nuclear genes and other molecules in the cell (particularly proteins), we infer that different genes must be switched on and off in the different cell types. What we know of the mechanisms of gene regulation has shown that inactivation of genes sometimes occurs through the binding of molecules that block the "reading" enzymes from the pertinent parts of the DNA. So the question arises: Is there some modification of the DNA (perhaps an indissoluble coating with proteins) that permanently prevents differentiated cells from expressing some genes?

The previous failure of attempts to transplant nuclei from adult vertebrate cells suggested that the answer to this question is "Yes." Wilmut's approach to the problem focused on the different stages of the cell cycle. He starved the cells from which the nuclei were taken, forcing them into the rest phase. His success in nuclear transplantation can be understood in one of two ways: either as the result of synchrony between the cell cycle of the donor from which the nucleus is drawn and the recipient, the enucleated oocyte into which it is inserted; or because the biochemistry of the rest phase makes the DNA of the transplanted nucleus susceptible to reprogramming (perhaps by weakening the bonds between DNA and the protein "blockers"). Whichever of these explanations is correct, the answer to the old question is now more complex. Differentiated cells do modify the genome in ways that make some genes inaccessible at some stages of the cell cycle, but at the rest phase they aren't so modified. This answer is a contribution to the project of understanding the dynamics of differentiation, and

2. The original article is I. Wilmut et al., "Viable Offspring Derived from Fetal and Adult Mammalian Cells," *Nature*, 385, 27 Feb. 1997, 810–813. The commentary is Colin Stewart, "An Udder Way of Making Lambs," *Nature*, 385, 27 Feb. 1997, 769–771.

thus of development as a whole, and it also points towards further projects (What is the exact character of the modification, and how is it undone?).

This example, or any of a thousand like it, can help us see the shortcomings of traditional ideas about the aims of the sciences. Nobody should be beguiled by the idea that the aim of inquiry is merely to discover truth, for, as numerous philosophers have recognized, there are vast numbers of true statements it would be utterly pointless to ascertain. The sciences are surely directed at finding *significant* truths. But what exactly are these?

One possible answer makes significance explicitly relative—the significant truths for a person are just those the knowledge of which would increase the chance she would attain her practical goals. Or you could try to avoid relativization by focusing on truths that would be pertinent to anyone's projects—the significant truths are those the knowledge of which would increase anyone's chance of attaining practical goals.

Neither of these is at all plausible as a full account of scientific significance, and the deficiency isn't just a result of the fact that both are obviously rough and preliminary. Linking significance to practical projects ignores areas of inquiry in which the results have little bearing on everyday concerns, fields like cosmology and paleontology. Moreover, even truths that do facilitate practical projects often derive significance from a different quarter. Surely the principles of thermodynamics would be worth knowing whether or not they helped us to build pumps and engines (and thereby attain further goals). Besides the notion of practical significance, captured perhaps in a preliminary way by the rough definitions given above, we need a conception of "theoretical" or "epistemic" significance that will mark out those truths the knowledge of which is intrinsically valuable.

Prominent efforts to understand the notion of epistemic significance, embodied in the writings of philosophers during the last three centuries and in the rhetoric of public paeans to scientific inquiry, attempt to show that inquiry is directed towards discovering a particular kind of truth, a kind scientists seek at all times, whatever practical projects they (or their contemporaries) may favor. The disciplines we pick out as sciences count as part of *science* because they aim at, and sometimes deliver, truths of this special kind, and they can be distinguished from technology precisely because the latter is focused on the practical. An allegedly context-independent notion of epistemic significance insulates science, or "basic science" at least, from social and moral values, by claiming that the achievement of epistemically significant truth is valuable in principle—even though, in actuality, that value might be compromised by ways in which the recognition of some truths would generate unfortunate consequences. Because I believe no such conception can be found, I take moral and social values to be intrinsic to the practice of the sciences.

My argument for this view doesn't depend on abandoning the idea that the sciences yield truth about nature or on giving up the ideal of objectivity; chapters 2 and 3 distinguish my position from one of the unacceptable images with which we began. Instead, I think there's a serious problem with traditional ideas about scientific significance, more specifically about epistemic significance. A gallery of bad pictures has held us captive. Dolly, the Scottish sheep, can help us recognize what is wrong with the traditional views and can point us in a better direction.

Traditional approaches suppose the notion of epistemic significance has nothing to do with us and our ephemeral practical concerns, and everything to do with the structure of the world. There are various ways to try to articulate the point. It is easy to start with muddy metaphors: some questions are "on Nature's agenda"; inquiry aims to discover "how Nature works." Personification is, however, hardly pellucid.

From the early modern period to the present, scientists and philosophers have tried to do better by invoking one of a family of interlinked concepts. So there arise the following well-known proposals:

> The (epistemic) aim of science is to achieve objective understanding through the provision of explanations.
> The (epistemic) aim of science is to identify the laws of nature.
> The (epistemic) aim of science is to arrive at a unified picture of nature.
> The (epistemic) aim of science is to discover the fundamental causal processes at work in nature.

Many thinkers have accepted more than one of these theses because they have recognized conceptual connections among the crucial terms ("explanation consists in subsuming the phenomena under laws," "explanation consists in identifying the fundamental causal processes," "laws of nature are those generalizations that figure in a unified account of the phenomena," "the fundamental causes are those described by the most general principles of a unified account of nature," and so forth). As we shall see later, it is not entirely clear that any of these grand conceptions will enable us to understand the hoopla about Dolly, but, for the moment, let us suspend worries that the particularities of her birth fail to live up to the large advertisements.

For our purposes, the important issue is not one that has figured in the large majority of worries about how we might come to apply the difficult concepts of law, cause, and explanation, but the question of which, if any, of the formulations of the last paragraph might identify an epistemic aim of science whose value could be convincingly defended. With respect to any of these projected achievements, it's appropriate for us to inquire, "What would be so valuable about gaining that?" There are immediate difficulties with the last two formulations. A unified picture of the world isn't something that wears its worth on its face—the question of why unity is so wonderful remains open. The commendation of causal knowledge does a bit better. Such knowledge plainly facilitates intervention in the world. Practical concerns are, however, not pertinent when we're out to fathom epistemic significance, and, when we bar them, there is again an open question about why knowledge of fundamental causal processes should be valuable.

Turn next to the suggestion that science is the search for natural laws, a proposal underlying many influential discussions of the sciences in the last three centuries. For the present, let's grant that a satisfactory account of natural law can be given, one that will distinguish genuine laws from accidental regularities. We can still ask why it's valuable to identify true statements with these special features.

Some major figures from the history of modern science would have answered this question by supposing that talk of laws is more than a bad pun: laws of nature are prescribed to the Creation by the ultimate sovereign, the Creator, and the world must conform to them. To seek the laws of nature is thus to reveal the divine rulebook, and to rejoice in the wisdom and beneficence of God. Copernicus, Kepler, Descartes, Boyle, and Newton all sounded the theme, and Newton's theological justification of his physics in a letter to Richard Bentley is typical: "When I wrote my treatise about our system, I had an eye upon such principles as might work with considering men, for the belief of a Deity; and nothing can rejoice me more than to find it useful for that purpose."[3] Similar ideas of a divine lawmaker whose statutes, once revealed, will inspire our admiration, resound throughout the eighteenth century and into the nineteenth (especially in Britain, where they are prominent in the Bridgewater treatises).

So here's *an* answer to the question I have posed. Knowledge of God ought to be our highest concern; disclosure of God's laws will promote this knowledge, thereby enabling us to "think God's thoughts after him"; what can be a more worthy goal than that? I doubt, however, that this will seem particularly persuasive. In light of our increasing knowledge of the history of the cosmos and of life on our planet, even committed theists are unlikely to feel that divine wisdom and beneficence are manifest in the creation: it all seems a curiously roundabout, baroque, inelegant, wasteful, and savage way of doing things.

In the twentieth century (and even in the nineteenth) the dominant articulation of the view that science aims to disclose the laws of nature has been thoroughly secular. Of course, even if there is no Creator and no divine rulebook, the universe might still be organized *as if* there had been a Creator with a rule-

3. Newton, *Opera*, vol. 4, 429. Quoted in E. A. Burtt, *Metaphysical Foundations of Modern Physical Science* (London: Routledge and Kegan Paul, 1924), 285. Burtt's work also contains excellent examples of similar ideas in the works of Copernicus, Kepler, Descartes, and Boyle.

book, but the secular surrogate loses the immediacy of the explanation of epistemic value. Recognizing the rules of organization might assist us with practical projects, but these concerns are irrelevant when we are trying to fathom epistemic significance.

The best way to develop the traditional approach is to appeal, at this point, to the idea of objective understanding and its correlate, objective explanation. Some truths are significant because they enable us to explain nature. Now in one very obvious sense, explanation is an activity provoked by actual or anticipated questions that arise in particular contexts, and explanations are directed at satisfying an envisaged audience that poses these questions. No defender of objective explanation should question this elementary point. Rather, what is claimed is that objective understanding consists in recognizing special relationships among the facts or events, so the criterion of success for an explanatory episode is the generation of this recognition, not any subjective satisfaction that the person given the explanation may feel. The sciences supply an *explanatory store* from which information or arguments can be drawn and adapted in the particular contexts in which understanding is sought and explanations given, and this store contains fully specific delineations of the relationships on which understanding depends. To put the point in its simplest terms, there is something which science supplies that provides an all-purpose basis for the practice of giving explanations: whatever interests people may have, whatever feelings of satisfaction or puzzlement they harbor, there is a set of relationships that, ideally, science presents to us, and the presentation brings a distinctive epistemic benefit.

One thought about those relationships is that they are revealed by showing how individual occurrences and states of affairs fall under general principles—explanation is a matter of subsumption under laws. Another thought is that they are recognized by seeing how the diverse phenomena of the natural world are integrated within a unified picture. Yet a third is that the relationships are causal, and that we appreciate them when we can identify the fundamental causes at work. But an important part of the view must be that the store is somehow systematic. It will fail as an all-purpose explanatory device if it is simply a long list of potential explanations, one for each context in which the desire for understanding might arise. Were that to be so, there would be no basis for a distinction between the epistemically significant and the epistemically insignificant—for every truth about the world would surely figure somewhere on the list, in the quite particular explanation that accounted for it, and, almost certainly, in giving explanations of particularities that flow from it.

The traditional search for a context-independent conception of epistemic significance is thus committed to the idea of a systematic organization of the truths about nature from which objective explanations may be drawn. I now want to scrutinize this commitment, and it will be useful to begin with a view that presented it most forthrightly.

The Unity-of-Science movement drew inspiration from examples in which particular scientific achievements were exhibited as derivative from others: Galileo's law of free fall and Kepler's laws of planetary motion could be viewed as consequences of Newton's gravitational theory; the laws of ideal gases could be incorporated first into the kinetic theory of heat and subsequently into statistical mechanics; thanks to the recognition that chemical bonds involve transfer or sharing of electrons, ascriptions of valence properties and, in consequence, laws of chemical combination, could be derived ("in principle") from atomic physics, ultimately perhaps from the basic equations of quantum mechanics. Extrapolating from these instances, it was proposed that all laws of chemistry could be derived from principles of physics, that all laws of biology could be derived from principles of physics and chemistry, that all laws of psychology could be derived from principles of biology (most notably neurobiology), chemistry, and physics, and so forth. Proponents of the unity of science understood quite clearly that the different sciences used special vocabularies, so that the envisaged derivations would depend on coordinative definitions that would link these vocabularies, and they pointed to the kinetic-theoretic identification of temperature as mean molecular kinetic energy as an example of the kinds of definitions that would ultimately be provided.

Attractive as the view may be, it has suffered from scrutiny of crucial junctions, most particularly those between the physical sciences and biology and between biology and psychology. Major difficulties have emerged. First, the successes achieved in the motivating examples seem to depend on the fact that the theories *reduced* (those exhibited as consequences of more fundamental parts of science) were individual laws or small collections of laws. Nobody has even the faintest idea what it would be to present biology or psychology as a cluster of laws. There are serious doubts concerning whether these sciences contain *any* genuine laws, and it is uncontroversial that there are highly significant parts of them that are not simply collections of laws: Dolly points to no general law of ovine (let alone mammalian) development.

Second, both biology and psychology seem to employ concepts that are not definable in the terms of the sciences proposed as reducing them. Defenders of the autonomy of psychology have pointed out how unlikely it is that there is a single characterization in terms of physics-plus-chemistry-plus-biology of the psychological state of thinking about the Unity-of-Science view (say), for the neural realizations and the underlying physicochemical conditions are very likely to vary from person to person. The situation is even clearer in the case of genetics. Nobody currently knows how to achieve a specification of the concept of gene in physicochemical terms: more pedantically, it is hard to see how to complete the schematic sentence, "*x* is a gene if and only if *x* is . . . ," by filling the blank with a structural description that will enable us to apply laws of physics and chemistry to derive conclusions about the behavior of genes. To be sure, an important necessary condition on genes is that they be segments of DNA or RNA; but of course there are lots of segments of DNA and RNA (most of them, in fact) that are not genes. The task is thus to identify the property that distinguishes the right segments of nucleic acid from the wrong ones. Contemporary efforts to discern the genes in reams of sequence data would be greatly aided if some such structural description were available, but, as molecular geneticists know all too well, the best they can do is to look for "Open Reading Frames"—stretches of DNA that show a relatively long interval between a codon for the initiation of transcription and a stop codon. (Computer searches pick out the ORFs, and investigators then use functional criteria to decide if new ORFs are genuine genes.) In addition, it's sometimes plausible to count regulatory "regions" as regulatory genes, to consider other nontranscribed regions as genes that have lost crucial parts of the regulatory apparatus, to see disconnected regions as parts of the "same gene," to identify overlapping genes or even genes within genes. Molecular genetics tells us at a staggering rate about the chemical structures of *individual* genes, but fails to provide a *general* specification. Matters are even worse when we move away from genetics and consider such important biological notions as *cell, organism, Drosophila melanogaster, species,* and *predator.*

The actual deliverances of the sciences accord rather badly with the Unity-of-Science view. But I now want to press a deeper point, scrutinizing the commitment to the provision of understanding through incorporation within a single overarching framework. Consider a fundamental Mendelian regularity which I'll formulate roughly as follows: genes sufficiently far apart on the same chromosome or on different chromosomes assort independently. How do we explain why this regularity obtains? The unity of science view sees us as gaining "objective understanding" from a physico-chemical specification of the notions of gene and chromosome and a derivation employing laws of physics and chemistry to yield the result about independent assortment. But this is quite unpersuasive. Rather we understand the regularity—as objectively as you please—by recognizing that the transmission of genes to gametes is a process of pairing and separation. Homologous chromosomes are brought together at meiosis, they may exchange some genetic material (hence the qualification about being "sufficiently far apart"—segments too close have a higher probability of being transmitted together), and one member of each pair is then passed on to a gamete. We assimilate the transmission of genes to all kinds of other processes which involve bringing together pairs of similar things and selecting one from each pair. What's crucial is the form of these processes, not the material out of which the things are made. The regularity about genes would hold so long as they could sustain processes of this form, and, if that condition were met, it wouldn't matter if genes were segments of nucleic acids, proteins, or chunks of Swiss cheese.

To reinforce the point, consider the regularity discovered by Dr. John Arbuthnot in the early eighteenth century. Scrutinizing the record of births in London during the previous 82 years, Arbuthnot found that in each year a preponderance of the children born had been boys; in his terms, each year was a "male year." Why does this regularity hold? Proponents of the Unity-of-Science view can offer a recipe for the explanation, although they can't give the details. Start with the first year (1623); elaborate the physicochemical details of the first copulation-followed-by-pregnancy showing how it resulted in a child of a particular sex; continue in the same fashion for each pertinent pregnancy; add up the totals for male births and female births and compute the difference. It has now been shown why the first year was "male"; continue for all subsequent years.

Even if we had this "explanation" to hand, and could assimilate the details, it would still not advance our understanding. For it would not show that Arbuthnot's regularity was anything more than a gigantic coincidence. By contrast, we can already give a satisfying explanation by appealing to an insight of R. A. Fisher. In considering sex ratios from an evolutionary point of view, Fisher recognized that, in a population in which sex ratios depart from 1:1 at sexual maturity, there will be a selective advantage to a tendency to produce the underrepresented sex. It is easy to show from this that there should be a stable evolutionary equilibrium at which the sex ratio at sexual maturity is 1:1. In any species in which one sex is more vulnerable to early mortality than the other, this equilibrium will correspond to a state in which the sex ratio at birth is skewed in favor of the more vulnerable sex. Applying the analysis to our own species, in which boys are more likely than girls to die before reaching puberty, we find that the birth sex ratio ought to be 1.04:1 in favor of males—which is what Arbuthnot and his successors have observed. We now understand why, for a large population, all years are overwhelmingly likely to be male.

I've been opposing two commitments of the Unity-of-Science view, the claim that the sciences can be hierarchically unified, and the view that integration within a single unified framework is the essence of objective understanding. It would be natural to respond to the arguments by proposing that, while they may doom a particular way of articulating the traditional view that epistemic significance attaches to those truths that can figure in an explanatory system, that is not the approach that the tradition should have preferred. But this is too sanguine. Invoking an ideal of objective understanding, based on a single unified framework of laws, was no arbitrary extension of the basic commitment, but an elaboration of the idea that science supplies a structure that is a resource for "objective understanding," whatever our contingent interests. Appealing to the Unity of Science specifies the character of this systematic, all-purpose structure. If the Unity-of-Science view fails, we need a substitute.

That, you might suppose, is easy enough. An obvious suggestion is that the discovery of laws (or the identification of causal processes) really does advance our understanding, although not in the way that the Unity-of-Science view suggests. Instead of a single system within which all "objective" explanations are subsumed, we proceed piecemeal, gaining understanding of nature by recognizing the laws of nature that govern the phenomena or the causal processes at work in the phenomena. This seems a salutary development, but it invites an obvious question. What is meant by "the phenomena"? Either the traditionalist intends, in accordance with the original motivation of science's agenda as set by nature, to think of providing resources for explaining *all* phenomena, or what is at issue is the explanation of the phenomena *that we find in need of explanation*, a vision that brings our contingent and evolving interests into the picture. What must be shown then, is how to reconcile the idea of *some sort of system* of laws or causes with the considerations that doomed the Unity-of-Science view.

Waiving concerns about the omnipresence of laws (that will occupy us shortly), the critique of the ideal of unified science displayed areas of inquiry that classified overlapping parts of nature in distinctive ways and offered their own (locally) unifying frameworks in terms of these classificatory schemes. Genetics, for example, focuses on DNA molecules but groups them in ways that do not map neatly onto physicochemical classifications and approaches the transmission of genes in ways that connect with (for example) processes of pairing and separation. So any complete system of laws of nature will consist of a patchwork of *locally unified pieces*, sciences with their own schemes of classification, their own favored causal processes, and their own systematic ways of treating a cluster of phenomena.[4] When we think about scientific inquiry as responding to a relatively narrow range of explanatory projects, to wit the kinds of questions we find worth posing, there's little harm in conceiving of this type of patchwork. But when we drop the reference to ourselves and our concerns, I see no reason to think there's any manageable system at all. To put the worry bluntly, why should we suppose that the number of classificatory schemes and unified treatments for all nature's phenomena is *finite*? The Unity-of-Science view had a simple answer to that question, since it proposed that the classificatory schemes of all the sciences were, ultimately, one, but once we've admitted plurality there's no reason for thinking we can stop. To revert to the motivating analogy of the

4. Nancy Cartwright thinks of the sciences as offering us a patchwork of laws. I agree with the point about the patchwork but believe that she places too much emphasis on the notion of natural law—if only in reacting against it. See her book *The Dappled World* (Cambridge: Cambridge University Press, 1998).

last chapter, the Unity-of-Science view made it look as though there was a fundamental set of maps from which any map we might care to use could be constructed, and so gave content to the conception of the ideal atlas. Once we abandon that view, it looks as though all that may remain is a collection of charts that may proliferate indefinitely with our changing interests.

At this point, a defender of context-independent goals for inquiry can reply that there is an as yet unformulated notion of objective understanding that will serve. We may not see yet how to divide the class of true statements about the world into those that are epistemically significant and those that aren't, but this is an important research project for philosophy of science. Let me explain why I am skeptical.

Those who seek context-independent goals for inquiry should admit that the explanation-seeking questions people pose take many different forms: we ask, "How?," "What?," "How is it possible?," and, of course, "Why?" The search for a philosophical notion of objective explanation has focused on Why-questions, conceding, tacitly or explicitly, that the topics of such questions reflect changeable human interests. But it has presupposed that there's a certain kind of information or argument that ought to be supplied in response to any explanation-seeking Why-question. The idea, then, has been that if we identify the explanatory store with the collection of all complete answers to Why-questions whose topics are true, there will be some propositions that pervade these answers, and these propositions are the epistemically significant truths. More exactly, given any Why-question whose topic proposition is true, there will be a special relation—the *relevance* relation—that holds between the topic and the objective complete answer. This relevance relation is independent of time and context, and whatever topics interest people, inspiring them to ask "Why?," the objective answers (the things that bear the relevance relation to the topic propositions) will always contain members of a set of true statements, the epistemically significant truths. So, for example, when explanation is taken to consist in showing how phenomena fall under laws of nature, any objective answer will contain some statement of law, and the laws will be selected as the epistemically significant truths.[5]

Plainly it's a disaster for this approach if all sorts of humdrum truths turn out to be epistemically significant. Now for typical mundane truths, statements about the contents of my cluttered desktop for example, there are everyday explanations in which those truths figure; my failure to find some pieces of paper

5. For a careful and precise account of the ways in which explanations work in everyday contexts, see Bas van Fraassen, *The Scientific Image* (Oxford: Oxford University Press), chap. 5; the proposal to see explanation as lawlike subsumption is most extensively developed by C. G. Hempel in the title essay of *Aspects of Scientific Explanation* (New York: Free Press, 1965).

when I'm looking elsewhere is explained by noting that the things I sought are buried in a pile on my desk. So it looks as though any truth, however banal, will occur somewhere in the explanatory store, unless we are offered a filter that lets just the "pervasive" truths enter the class of the epistemically significant.

We can now see why the approach in terms of unified science was so promising, for it imagined that ideally complete explanations use the same fundamental principles again and again. Without recourse to the Unity-of-Science view, we have to look for some context-independent relevance relation that will generate the right filter. One natural suggestion is that explanations are given by furnishing causal information that bears on the topic. Now in everyday explanations the kinds of causal factors that people provide are heavily context-dependent: the lawyer, highway engineer, automobile mechanic, and psychologist may offer quite different accounts of why the Princess of Wales had a fatal accident. Perhaps, however, we should view the diverse appeals to causal factors as context-dependent selections from the *complete* cause, supposing that there's a context-independent relation between the topic (specifying the circumstances of Princess Diana's death) and a complete description of its causal antecedents.

But we can completely specify the causal factors that produced an effect at any given time prior to the effect, so that focusing on a particular time already involves a further selection. To avoid context-dependence, one must invoke the idea of a complete causal history, an imagined account that shows how the effect described by the topic occurred as a consequence of events in the remote past. The view, then, must be that the objective answers describe some vast causal history, and that these serve as a store for ordinary explanations by permitting selections that are attuned to the interests of the intended audiences.

This idea is vulnerable to two difficulties. First, recall one of the considerations that doomed the Unity-of-Science view. Neither the sequence of "male" years in London nor the independent assortment of genes is understood by grinding out the full causal details: a narrative drawn from the deep recesses of the past would fail to offer the type of information sought. We might honor the idea that here too explanation consists in the provision of causal information, but only by recognizing that it is a different type of causal information, one not captured in the allegedly ideal and complete causal history. We understand sex-ratios by seeing the state we wish to understand as an equilibrium and identifying the factors that maintain the equilibrium—in a sense a causal account, but one that doesn't relate effects to completely specified antecedent causes.

The second problem arises when we consider the goal at which the account of explanation aims. We worried earlier that all sorts of mundane truths would figure somewhere in the explanatory store. But with respect to virtually any truth, however humdrum, we can devise a complete causal narrative in which that truth plays an essential role. So how is the filtering to be done? Perhaps by supposing that the epistemically significant truths occur in *every* complete causal narrative. Given the difficulties of the Unity-of-Science view, it looks as though ideal explanations will describe causal processes that occur at different levels, but perhaps there are *some* truths common to all members of the explanatory store, for example descriptions of very early stages of the universe. To broaden the epistemically significant beyond early cosmology and particle physics, one might suggest that epistemically significant truths are those that occur in a very large number of the complete causal narratives. How do you count? Given the continuity of time, it looks as though any statement that occurs in a complete causal narrative figures in an infinite number of such narratives (indeed, continuum many), for it will be an essential part of the "objective explanations" of all those statements that describe "downstream" states and events. In terms of numerical frequency, all truths are on a par.

The enterprise thus strikes me as hopeless. For there's a general problem. Everyday explanations seem quite varied in their offerings of causal information (and maybe in other types of information as well). To pick out a context-independent relevance relation that covers all this diversity requires one to portray individual acts of explanation as selecting from much vaster entities, ideal explanations that have many constituent propositions. Just about any truth will turn up in the resultant store. Because of the deficiencies of the Unity-of-Science view, the statements one would like to pick out as epistemically significant will not be all-pervasive. So there will be no simple solution to the filtering problem. The only option seems to be to resort to counting, and this fails because the classes to be compared are all infinite.

We can free ourselves from this bind by developing a different approach to "objective explanation." Given a topic that is of interest to us and a relevance relation, the objective explanation is whatever complex of truths stands in the appropriate relation to the topic. Just as the topics of interest to us evolve in the growth of the sciences, so too with the relevance relations. Perhaps many of these are broadly causal (although we do recognize other types in seeking mathematical explanations, for example). Even the most cursory survey of our practices reveals the heterogeneity of the relevance relations that pertain to our questions: sometimes we are interested in triggering events, sometimes with enduring features that are taken to constitute the "natures" of the things under study, sometimes with the intentions of agents, sometimes with conditions that maintain an equilibrium, sometimes with factors that are to the advantage of an organism. Frequently, relevance relations reflect our interest in the covariation of properties we find salient or in factors that we can manipulate and control. Objective explanation goes on in the sciences, then, but only against the background of our questions and our interests. The most we can expect from a theory of explanation is some understanding of how these questions and interests shift as our inquiries, and the complex environments in which they occur, evolve.[6]

I now want to approach the issue in a different fashion, considering the ways in which judgments of significance are made in the everyday practice of the sciences. Whether one turns to the specialized journals of particular subdisciplines (*Physical Review*, *Cell*) or to the general journals in which publication is most difficult (*Science*, *Nature*), it's overwhelmingly obvious that new laws are very hard to find. Prominent articles tell us about the distribution of minerals in particular parts of the earth's crust, about the relative sizes of australopithecine skulls, about the sequences of the genomes of bacteria, worms, and flies, and, of course, about that celebrated sheep. Why should any of these studies be hailed as significant?

In all such instances, we can tell the same sort of story I summarized in the case of Dolly. There are broad questions we find interesting—What were our hominid ancestors like? How do single-celled organisms regulate their metabolism?—and we can see the findings as advancing the project of answering them. Often there are practical problems—of understanding earthquake zones or combating Lyme disease—on which the research bears. Indeed, in many cases though not in all, epistemic and practical interests are interwoven.

Defenders of science as the search for laws and objective explanations have an obvious strategy for responding to these examples. The goals of a vast and ambitious enterprise are not necessarily revealed in everyday activities, and a myopic focus on the brushwork will not reveal the splendor of the picture. So, they might contend, the very particular investigations I report are significant because they are small contributions to attaining the types of epistemic significance that the tradition celebrates: the laws and the descriptions of fundamental causes will emerge from them, perhaps in one of those rare articles published every few decades in which some fortunate scientist, standing on the shoulders of a pyramid of under-laborers, displays what the entire venture has been aiming towards all along.

There's a valuable point here, in the recognition that significance accrues to work that would strike many outsiders as arcane, because of the advancement of a much larger project. But I think the response errs in misunderstanding the

6. In an earlier essay, "Van Fraassen on Explanation," *Journal of Philosophy*, 84, 1987, 315–330, Wesley Salmon and I framed issues about explanation in terms of a choice between giving an "objective" account of explanation and a pluralism in which "anything goes." We overlooked an important possibility, that variable interests might promote different topics and different relevance relations. Not every relation counts as a relevance relation, but one can think of a family of relevance relations, bound together by loose resemblances, indefinitely extendable and coevolving with the history of inquiries and the social ventures they serve. This may have been van Fraassen's own position.

interrelations among pieces of scientific work, the channels along which scientific significance flows. It thinks in terms of a Baconian hierarchy: the contents of *Nature* and *Science* are pieces of information that will be systematized into general laws, and, ultimately, into overarching theories—significance runs (drips?) from the envisaged theoretical top to the mundane accomplishments at the bottom. The demise of the Unity-of-Science view ought to make us suspicious about parts of this, but the examples I have cited, with my brief explanations of how they integrate into larger projects, should inspire a quite different concern: is there any reason to think that significance flows from the general (or the "causally fundamental") to the particular, rather than having its source in very specific concerns about particular types of properties of entities that matter to us (the crust of our planet, our ancestors, bacteria that are human pathogens)? Rather, the connections that confer significance seem to radiate in many different directions, so that a map of an area of inquiry that reveals how its claims and projects earn their significance might look more like a tangled skein than a hierarchy. An elaboration of this view will show, I believe, how it does much greater justice to the way in which scientific significance attaches to the work scientists actually do.

Back to Dolly. In all epochs, and in all cultures, people have been struck by the fundamental phenomenon of development, the unfolding of characteristics from initially tiny fragments of organic material, with preservation of species-typical traits. This phenomenon must have been evident to those who first domesticated plants and animals (the collection and planting of seeds bears witness to an appreciation of it), and some aspects of it were apparent even in the case of our own species. Conceiving of reproduction as a process in which something is transmitted from parents to offspring, investigators since antiquity have struggled to learn what this something is and how it interacts with the rest of nature (particularly the bits of the world with which it comes into contact) to produce a new organism. The first question—"What is the hereditary material?"—obviously fuels the large questions of genetics, from Mendel's observations to the present. The second bifurcates into two kinds of issue: the description of the processes that lead from the first stages of the nascent organism to its adult form, and the fathoming of causal processes of particular interest to us. As we learn that the first stage of the new organism is a zygote (a fertilized egg) and that development proceeds via cell division, we pick out some causal processes for attention. Thus we ask, "How are genes activated and inactivated?" (because we think of the hereditary material as playing a role in guiding the organism towards a species-typical phenotype), "How do the major cellular movements that lay down the organism's body plan take place?" (because we see, for example, that all vertebrates have a common form we can trace to changes that take place in the formation of the notochord), "How do cells of different types differentiate?" (because we recognize physiological, and subsequently biochemical, differences among the cells from different bodily systems), and so forth. The last of these questions forms the backdrop to Wilmut's work, in just the ways I indicated earlier.

So we have found our way from a natural preoccupation of virtually all people to the birth of a lamb in a Scottish research station. We could trace similar paths from the initiating question for developmental biology to many other inquiries—efforts to map and sequence the genome of the soil amoeba *Dictyostelium discoideum*, attempts to breed mutant zebrafish of particular types, computer programs that try to simulate the growth of a chick forelimb, biochemical assays of tissues in *Drosophila*, mathematical analyses of snail shell patterns. Instead of automatically assuming that these efforts, to which highly intelligent and extensively trained people devote large portions of their lives, are directed at some enterprise of great generality, we can actually explore why they have come to be at the focus of inquiry, recognizing the affiliations to practical projects as well as to large questions that naturally excite human curiosity.

As with the example of map-making considered in the last chapter, where we saw that the kinds of maps we construct are shaped by evolving interests, so too the questions we take to be significant and the endeavors we pursue in attempts to answer them co-evolve with all sorts of practical projects. Fields of science are associated with structures I shall call *significance graphs* that embody the ways in which their constituent research projects obtain significance. A significance graph is constructed by drawing a directed graph with arrows linking expressions, some of which formulate questions that workers in the field address, others that encapsulate claims they make, yet others that refer to pieces of equipment, techniques, or parts of the natural world (figures 1, 2). The significance graph reveals how to explain the significance of various items—where 'item' is an all-purpose term for questions, answers, hypotheses, apparatus, methods, and so forth. One would account for the significance of the item to which the arrow points in terms of the significance of the item from which it comes. Arrows thus display the inheritance of scientific significance.

In talking of the "explanation of significance," I intend to make explicit what workers in the field know at the time to which the net is indexed. The commentary on Dolly told the broader public what researchers typically take for granted. As a field grows, however, the character of the significance graph changes so that later explanations of significance are quite different from those that would have been given earlier. Moreover, the significance of an item may well be overdetermined, and researchers with different interests may give priority to alternative linkages. Some prize Dolly because of the possibilities she represents for improving livestock, others because she contributes to our understanding of cellular differentiation. We can adopt a *field-centered* perspective on significance graphs, one that shows how significance is inherited within a par-

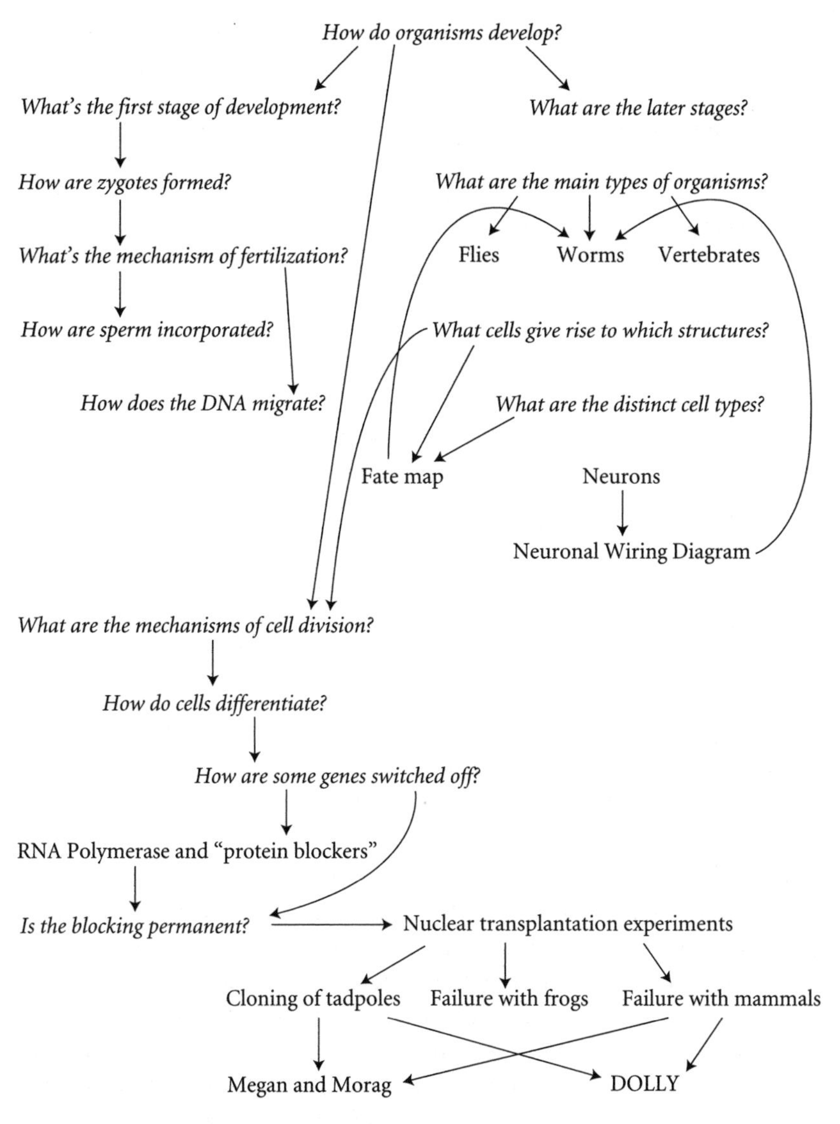

Figure 1. An extremely partial depiction of the significance graph for developmental biology.

ticular area of research, viewing Dolly solely within the purview of developmental biology (say). Or we can take an *item-centered* perspective, looking at all the ways in which a particular node in the significance graph, the one designating Dolly for example, gains significance for science. The perspectives are compatible and valuable for different purposes.

One principal difference between thinking in terms of significance graphs and the more traditional conception of the sciences as seeking laws is a far more

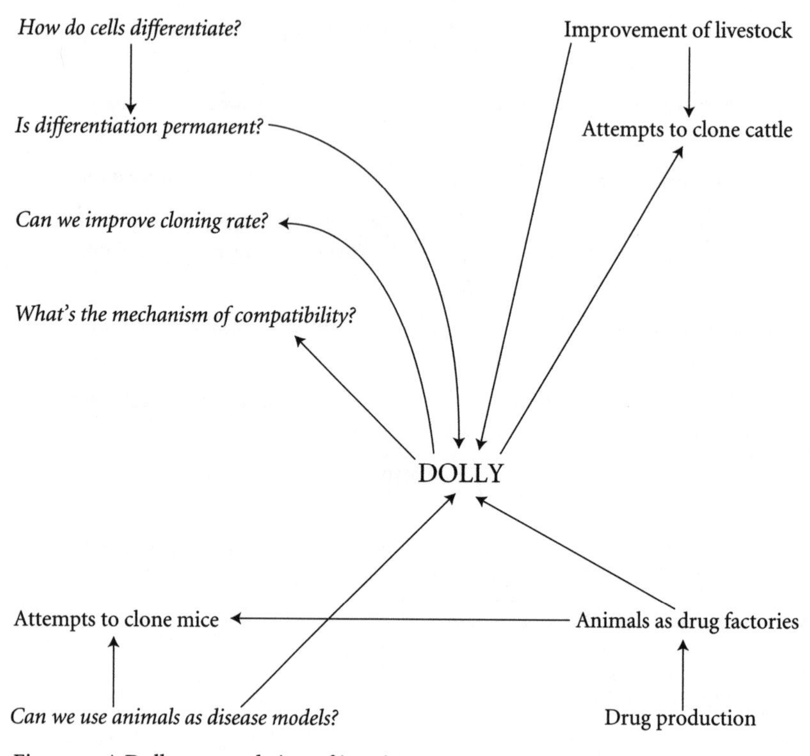

Figure 2. A Dolly-centered view of inquiry.

pragmatic approach to generality. Large generalizations are good where we can find them, for they enable us to fit phenomena into a broader framework, to explain more, to predict more, and maybe to intervene more successfully. There may be some areas of scientific inquiry where we can achieve precise, accurate, generalizations of large scope—the parts of physics that have inspired many philosophers may be like this—but the most common aspect of *la condition scientifique* seems to be that we have to make compromises among generality, precision, and accuracy.[7] Hence significance graphs do not embody the idea that significance (or epistemic significance) is always a matter of achieving, or pointing to, universal laws.

What, then, is the ultimate source of epistemic significance? The answer is, I think, commonplace and disappointing to those who expect a grand theory that will invest the sciences with overriding importance. Recall our explanation of Dolly's significance. It began from the idea that we wanted to understand

7. The point was formulated beautifully by Richard Levins in *Evolution in Changing Environments* (Princeton: Princeton University Press, 1966). For a deep skepticism about the possibility of finding precise generalizations even in physics, see Cartwright, *The Dappled World.*

how an organism's characteristics unfold from a tiny piece of organic material. If someone asked why we want to understand that—or why we want to know why the heavenly bodies move as they do, or why we are interested in the evolution of our hominid ancestors—it would be hard to say very much. We expect other people to see the point of such questions, and we describe those who don't as lacking in "natural curiosity." Partly as the result of our having the capacities we do, partly because of the cultures in which we develop, some aspects of nature strike us as particularly salient or surprising. In consequence we pose broad questions, and epistemic significance flows into the sciences from these.

Human beings vary, of course, with respect to the ways in which they express surprise and curiosity. Some are disposed to ask more, others less. Typically, we respond to the diversity with tolerance, explaining some of the variation in terms of differences in cultural or educational context. But tolerance has its limits, and we do count some of our fellows as pathological, either because they obsess about trifles or because they are completely dull. In claiming that the sciences ultimately obtain their epistemic significance from the broad questions that express natural human curiosity, I am drawing on this practice of limited tolerance, on our conception of "healthy curiosity" and the commonplace thought that most of us, given minimal explanation, would find interesting the global questions that stand at the peripheries of significance graphs.

Significance graphs evolve. As information accumulates, new connections are forged, new practical projects are designed, new questions emerge as tractable ways of pursuing inquiries already established. Precisely because the same entities sometimes serve as resources both for important practical ventures and for theoretical work, epistemic significance today may bear the traces of yesterday's practical significance. Because some instruments, techniques, sites, or model organisms become embedded in the significance graphs of different fields, so that researchers know how to use them, the evolution often shows a kind of inertia. Alternative choices made earlier would have led to a different development of the field, so that, in quite particular ways, the development of the sciences is thoroughly contingent.[8]

As our inquiries evolve and different phenomena become salient for us, we introduce new classifications, dividing up the world in novel ways. Sometimes we arrive at new views about which entities are single objects (think of the recognition that lichens consist of symbiotic pairs of organisms); more frequently, we group old entities together in different ways. Our understanding of objects and kinds of things evolves with our significance graphs, driven by changing expressions of natural curiosity and by our practical needs. Further-

8. For articulation of the point in the context of experimental physics, see Peter Galison, *Image and Logic* (Chicago: University of Chicago Press, 1997). The dominance of certain organisms—for example, *Drosophila*—in contemporary biology is extremely obvious.

more, the contingent decisions made today modify the pressures under which the graphs evolve, for the phenomena into which we inquire sometimes change in response to our activities. One obvious example comes from twentieth-century biomedical research, in which one generation's struggle against what it views as the most devastating diseases can open up niches for disease vectors, thereby allowing the evolution of new pathogens. If pessimistic forecasts about infectious diseases are even partly justified, the medical problems our descendants are likely to confront will be caused by microorganisms whose existence depends on our ancestors' choices.

I have tried to outline a view of scientific significance that is very different, both in its character and its consequences, from that which has dominated traditional reflections on inquiry. If I am right, then the analogy of the last chapter has been vindicated. Like maps, scientific theories—or, better, significance graphs—reflect the concerns of the age. There is no ideal atlas, no compendium of laws or "objective explanations" at which inquiry aims. Further, the challenges of the present, theoretical and practical, and even the world to be mapped or understood, are shaped by the decisions made in the past. The trail of history lies over all.

II

THE CLAIMS OF DEMOCRACY

SEVEN

The Myth of Purity

CONSIDER A STANDARD DEFENSE of unpopular scientific research. Moral, social, and political concerns, it's frequently suggested, should not be invoked in the appraisal of investigations. The sciences seek to establish truths about nature. How the resultant knowledge is used is a matter for moral, social, and political debate, but it is intrinsically valuable for us to gain knowledge. If the circumstances in which knowledge is applied are likely to generate harmful consequences, then that is a sign of defects in the social milieu that surrounds the sciences, and, ideally, we should try to gain the knowledge and remove the defects.

As I remarked in chapter 1, everyone will agree to qualify this defense in one way. Investigations involving procedures that would violate the rights of subjects (systematic torturing of neonates to measure their capacity for pain, for example) are properly rejected on moral grounds. Critics of the defense often seek further concessions on two grounds, first by supposing that moral, social, and political values affect decisions about which projects are worth pursuing, and, second, by claiming that such values partially determine which statements are accepted as "true."

In the preceding chapters, I have been disentangling the criticisms. There is no need to abandon the everyday conception that inquiry yields truth about independent objects. Nor should we suppose that ideals of objectivity are misguided and that, because of rampant underdetermination, scientific decisions are made, perforce, by invoking moral, social, and political values. Yet my account of the ways in which our evolving interests draw new boundaries in nature and of how the concept of scientific significance reflects contingent interests enables us to see how to develop the first criticism. The standard defense of the last paragraph depends on a view of the aims of the sciences we ought to abandon.

All kinds of considerations, including moral, social, and political ideals, figure in judgments about scientific significance, and hence in the evolution of significance graphs. Inquiries that appeal to us today, and that we characterize as epistemically significant, sometimes do so because of the practical projects our predecessors pursued in the past. With our eyes focused on the present, it's easy to deny that these inquiries are in any way connected with broader values. A longer view would reveal that the questions we pose, the apparatus we employ, the categories that frame our investigations, even the objects we probe, are as they are because of the moral, social, and political ideals of our predecessors.

It's hard not to sympathize with the physical chemist who dismisses the idea that his research is suffused with the values of bourgeois males of European descent, and who bluntly declares that he's just out to analyze (or synthesize) the molecules. When the critique is directed against modest realist claims, that is when it's suggested that the values in question are reflected in the structures the chemist presents in his research reports, then the charge lapses into absurdity. Yet when we formulate the worry as one about scientific significance, matters are different. Why have those molecules been selected for analysis or synthesis? Or, in some instances, why do those molecules exist at all? A significant number of contemporary investigations go forward because entrepreneurs believe that studying *just these* molecules will help increase their profits. Even when such direct links are absent, however, some lines of chemical inquiry take the form they do because of the practical decisions of earlier generations. It seemed morally legitimate to previous researchers to find ways of ameliorating the debilitating effects on workers of the hazardous environments in which they labored, and so to focus on the problem of understanding certain complex molecules and their interactions. Some generations later, the chemist wrestles with the theoretical problem of fathoming a molecular structure without any conception of the filiations that connect his research with a past policy of "protecting" a group of workers, or with the future applications to which his findings may give rise.

We need to scrutinize the myth of purity. The most popular form of the myth supposes there is a straightforward distinction between pure and applied science, or between "basic research" and technology. I shall try to show that these divisions are not so simple.

Pure science isn't differentiated from applied science or technology by the sites at which it is practiced. Industrial laboratories contain "pure" researchers, and academic environments harbor people dedicated to technological ventures. Nor can we make a separation in terms of products. Basic science produces devices as well as knowledge, and technology sometimes yields knowledge as well as devices—indeed it's tempting to argue that the manufacture of a device inevitably brings with it new knowledge about the ways in which parts of nature can be harnessed to work together. Instead of looking at external signs, like places or products, we do better to try to mark the distinction in terms of aims. The aim of science (pure science, basic research) is to find truth; the aim of technology (applied research) is to solve practical problems.

There is surely something right about this, but it will not do as it stands. The aim of science is not to discover any old truth but to discover significant truths. Recognizing the distinction between epistemic and practical significance, we might propose that pure science aims to find those truths whose only significance is epistemic. Yet this is vulnerable to the possibility that inquiries undertaken solely for the sake of satisfying curiosity might turn out to have practical payoffs, and would thus be debarred from counting as parts of pure science.

We do better to deploy the notion of aims in its most natural home, referring to the aims of individual agents rather than those of some abstraction (like science or technology). Let's say, then, that an investigator is practicing pure science just in case the investigator's aim is to address a question in the significance graph solely because of the epistemic significance that that question inherits. We can explain what this scientist does simply by adverting to the epistemic significance that would come from her success and seeing her as motivated by her perception of this significance. She wants to find some elusive particle, say, solely because she sees that the discovery of this particle would answer theoretical questions about the structure of matter; whether its discovery would have any practical implications is of no concern to her. Her technologist colleagues, by contrast, do the things they do solely with the intent of resolving practical problems and have no interest in whatever epistemic significance may accrue to the truths they discover.

There are obvious complications. The pure scientist we've envisaged is extraordinarily high-minded. Considerations of fame or fortune (or grant renewal) are no part of her motivation. When such personal motives are present, how should they be classified? Are there virtually always practical concerns hovering in the background, if only in the conscientious researcher's concern to give satisfaction to employers or funding agencies? Perhaps the simplest response is to suppose that these kinds of motivations occur equally in pure scientists and in those who practice technology, so that they can be ignored for the purposes of drawing the distinction. In any event, I shall henceforth ignore them.

Deeper difficulties come from the multifarious interconnections of the epistemic and the practical in significance graphs. Dolly's significance derives in part from connections to broad issues in development, in part from her agricultural and medical promise. After Dolly, investigators may undertake ventures in nuclear transfer using different donor cells in different mammals. Their inquiries satisfy curiosity—Are some mammals easier to clone than others? Are some cells especially good for supplying nuclei?—but they may also advance practical projects. How should mixed inquiries be classified?

One response is to interpose a third category between science and technology: there's basic science (the pursuit only of epistemic significance), applied science (with both epistemic and practical significance), and technology (only of practical significance).[1] Ventures in mammalian cloning would be taken to belong to applied science. This, however, seems doubly unsatisfactory, for it lumps together the investigator who is out to fathom the molecular changes involved in cellular differentiation and the researcher who wants to find reliable methods of breeding superior livestock. Once again, we need to attend to the particular intentions of the scientists involved. There's no all-purpose tag that can be pinned on particular constituents of significance graphs.

With respect to cloning, it's easy to envisage two extreme cases and to classify them by attending to the researchers' aims. Yet an inquirer's motives can genuinely be mixed. Would-be cloners might want both to address broad questions about development and to produce a better pig. In such cases, the intermediate category of "applied scientist" (or something similar) seems an attractive idea, exposing the double nature of the lines of connection in the significance graph. On further reflection, we should appreciate that not all that is mixed is mixed equally. In the middle sits an investigator equally devoted to embryological insight and porcine perfection. Just to her technological right is a colleague who gives slightly greater weight to bringing home the bacon, while just to her scientific left is another colleague whose priorities are the reverse. Indeed, we can envisage a chain of researchers extending from the pure embryologist at the one end to the animal breeder at the other. Where along this chain do we want to mark the boundaries of however many categories we propose to introduce?

An appropriate answer to that question would point out that, despite the difficulty of fixing transition points, it may still be valuable to distinguish the extremes. Even though we can't find a sharp distinction between "pure science" and technology, we can still use a vague distinction that separates certain very clear cases—the imaginary embryologist and animal breeder, for example. We understand the easy cases and the hard cases by disclosing the structures of the pertinent significance graphs and the ways in which different investigators respond to those structures and seek to extend the nets. "Pure science" is what

1. An approach along these lines is developed by Ilkka Niiniluoto, "The Aim and Structure of Applied Research," *Erkenntnis*, 38, 1993, 1–21. Niiniluoto is firmly in the tradition of seeking context-independent aims for inquiry, but he offers an account of the science–technology distinction that is unusually sensitive to some of the hard cases. Another important attempt to fashion an intermediate category is offered by Donald Stokes in *Pasteur's Quadrant* (Washington D.C.: Brookings Institute Press, 1997). Stokes's discussion demonstrates quite convincingly that the motives of tackling a practical problem and contributing to "basic science" aren't incompatible.

pure scientists do, and pure scientists are those people whose research is guided by the lines along which epistemic significance flows.

There are two further complications, however, both prefigured in earlier parts of this discussion. First, we may look backward and recognize that the historical explanation for the current epistemic significance accruing to a line of inquiry turns in part on some practical project from the past. Second, we may look forward and recognize that there are readily envisageable ways of linking the results of the inquiry (or the possible results if the inquiry develops in a particular foreseeable way) to practical projects that others could be expected to pursue. If research is to be genuinely pure, how should the investigator's aims accommodate these filiations to the practical?

There are natural answers. A chemist, working on a molecule of current "theoretical" interest, may well not know or care that the molecule came to scientific attention because of past efforts to find a cheap way to appease the public about conditions in mines. Unlike others who work on the molecule because it will speed up a commercially important industrial process (albeit at an environmental cost), our chemist has no ties to entrepreneurs and no concern for the practical applications. As he never tires of explaining, he simply finds the problem of figuring out the structure a fascinating challenge. Since his aims are only to achieve results of epistemic significance, he is a pure researcher. Or is he?

Once again, we can contrast extreme cases. When any links to practical projects are buried in a distant past, with no bearing on contemporary applications, and when it's very hard to forecast how results from this inquiry could be adapted to solve practical problems, then researchers can quite legitimately declare their intentions to be thoroughly epistemic. However, when only a little curiosity is needed to see that the current significance graph has been shaped by dubious ventures from the past, or when the propensity of others to engage in morally consequential applications ought to be obvious, the researcher who proclaims solely epistemic intent is guilty of self-deception (at the very least). Tom Lehrer made the point in a witty lyric:

> "When the rockets go up, who cares where they come down?
> That's not my department," says Werner von Braun.

Pure researchers, then, are not simply those whose intentions are entirely to promote epistemic significance but whose lack of interest in the practical can be justified.

We've been considering the complexities of the distinction between science and technology, and it's worth stepping back to remind ourselves of why the distinction has seemed so important. As I noted at the beginning of this chapter, the fundamental point seems to be to limit the scope of moral, social, and political appraisal. If a clear separation can be made, then the line of defense considered at the beginning of this chapter can be articulated. Beyond requiring that researchers pursue their experiments in morally appropriate ways (treating their experimental subjects properly, dealing honestly with fellow scientists, and so forth) there are no further moral, social, and political standards to which the practice of science is accountable. Such standards arise only in the context of applied science or of technology.

The myth of purity proposes that there is a distinction that fulfills these purposes. The considerations of this chapter oppose the myth. We may be able to identify certain people as practicing "pure research," but our classification depends not only on their intentions but also on whether those intentions can be justified. In other words, insofar as the distinction between pure science and technology can be drawn, it depends upon a *prior* judgment to which moral considerations are pertinent. The claim that a particular inquiry is a piece of pure science can only be evaluated in light of the character of the significance graph, the intentions of the investigator(s), and the possibility of justifying a practice of ignoring any connections to practical concerns. Very frequently, the complex intertwining of the epistemic and the practical and the mixed motivations of actual researchers will make the application of any simple distinction (or set of distinctions) impossible, but, even when we separate out these complications, the links to past projects and to future possibilities have to be assessed *before* we can count the inquiry as a piece of pure science. Flourishing the badge of purity isn't automatic. The label has to be earned.

I'll conclude the discussion by illustrating my point with one of the most obvious examples of pure research, one that may initially seem to vitiate many points of the past two chapters. At the frontiers of contemporary theoretical particle physics, researchers explore extremely abstract mathematics in trying to find a unified account of fundamental forces and the elementary constituents of matter. Surely in this instance the line of thought with which we started the chapter seems to work: practical consequences, for good or ill, are too remote to be specified; rather the inquiries are pursued because of the value of uncovering the deepest ("most beautiful") truths about our universe.

Let's accept the claim that practical consequences are indeed remote, that, unlike the comparable situation at earlier stages of atomic physics, there are no relatively straightforward ways to try to deploy principles and theories that are likely to emerge from the investigation in order to generate vast amounts of energy. We can still ask why the project is assigned such high value. With the demise of the Unity-of-Science view, the answer can't be that we're going to arrive at a theory from which all other parts of science are destined to flow. Rather the significance of the work lies in the interest for us of identifying the ultimate constituents of matter. At various stages in past inquiry, attempts to answer that question have been connected to all sorts of practical concerns, but, even if we set those to one side, there's a fundamental point about the justification of further inquiry. To concentrate on the epistemic significance of a unified treatment of gravitation and the other three "fundamental forces" is to presuppose a judgment about the relative value of answering a particular set of questions in mathematical physics and alternative ways of extending the collection of significance graphs that the current generation of researchers has inherited from the past. There are any number of ways in which we might go on from where we are (including some that would revoke past decisions), and the resources of equipment, time, and talent are finite.[2] Engaging in research that does no foreseeable harm may be unjustified because of the good that the researchers who carry it out fail to do. Once we have abandoned the idea of a context-independent conception of epistemic significance, we see that judgments about lines of inquiry inevitably weigh the rival merits of various practical goals and various ways of satisfying human curiosity. This applies to the "purest" cases just as it does to the areas of science that are obviously intertwined with applications.

None of this is to suggest either that attempts at a theory of superstrings (or similar ventures at the theoretical reaches of physics) are impure or that they are unjustified. It seems to me eminently possible that researchers who undertake this project are motivated by concerns of epistemic significance alone and that they are entitled to ignore any practical linkages. But my hunch that their research is pure and well motivated depends on supposing that the results of a moral, social, and political appraisal would vindicate it. The myth of purity is the claim that gesturing at the absence of any practical intent is enough to isolate a branch of inquiry from moral, social, or political critique.

I shall elaborate upon this theme later. First, however, I want to consider an example in which an ideal of pure inquiry has been invoked to ward off political objections.

2. Similar points were made during the 1960s, at a time when there was serious debate about the foundations of issues in science policy. For a forthright statement, see Alvin M. Weinberg, "Criteria for Scientific Choice," in *Criteria for Scientific Development: Public Policy and National Goals*, ed. Edward Shils (Cambridge, Mass.: MIT Press, 1968).

EIGHT

Constraints on Free Inquiry

IN THE MID-1970S, A GROUP OF SCHOLARS, including prominent biologists as well as academics from other disciplines, wrote scathing indictments of conclusions they claimed to find in E. O. Wilson's much-lauded book, *Sociobiology: The New Synthesis.* Wilson had argued that a Darwinian analysis of human social behavior revealed that certain features of contemporary societies were deeply rooted in human nature, and thus unmodifiable by adjusting the environments in which people develop. In particular, he suggested that current sex roles are inevitable, that xenophobia cannot be eradicated, and that we can expect that any society will be based on intense competition that generates inequalities. The Sociobiology Study Group of Science for the People charged that these conclusions were both unwarranted and politically dangerous in their apparent support of reactionary policies. In two replies, Wilson disavowed many of the conclusions, claiming his critics had misinterpreted his book, and he ended each article by recalling a traditional liberal theme. The linking of explicitly political considerations to the scientific discussion was an instance, he averred, of "the kind of self-righteous vigilantism which not only produces falsehood but also unjustly hurts individuals and through that kind of intimidation diminishes the spirit of free inquiry and discussion crucial to the health of the intellectual community."[1] In a more expansive treatment, he closed a second reply with the following paragraph:

> All political proposals, radical and otherwise, should be seriously received and debated. But whatever direction we choose to take in the

1. E. O. Wilson, "For Sociobiology," *New York Review of Books,* 11 Dec. 1975; reprinted in Arthur Caplan, ed., *The Sociobiology Debate* (New York: Harper, 1978); see p. 268.

> future, social progress can only be enhanced, not impeded, by the deeper investigation of the genetic constraints of human nature, which will steadily replace rumor and folklore with testable knowledge. Nothing is to be gained by a dogmatic denial of the existence of the constraints or attempts to discourage public discussion of them. Knowledge humanely acquired and widely shared, related to human needs but kept free of political censorship, is the real science for the people.[2]

Wilson thus cast his critics as attacking precepts about the value of free inquiry that have a rich heritage in the liberal democratic tradition, and, consequently, are typically accepted without question.

Wilson's critics quickly disavowed the charge that they were trying to hedge free inquiry with political constraints.[3] Suppose, however, they had queried the traditional precepts, asking why the value of free inquiry should outweigh other moral, social, and political concerns. In that case, the debate would have turned to the tradition on which Wilson tacitly drew, probably to the writings of the most eloquent defender of free expression, John Stuart Mill, and to the second chapter of *On Liberty.*

There Mill advances four arguments in favor of free expression. He begins with the fallibility of human opinion, pointing out that even though we may feel certain of the truth of our beliefs we may still be mistaken, and claiming that it is important to guard against error by allowing open discussion of rival points of view. Secondly, he notes that views that are, as a whole, false, may contain some truth, even some truth that orthodoxy currently fails to recognize, so that free discussion may guide us to improved opinions. Furthermore, received beliefs that are not subject to criticism from alternative perspectives may come to be held dogmatically, "in the manner of a prejudice"; and, finally, Mill notes that the meaning of the doctrines may become lost. Now it's noteworthy that all of these considerations depend on an ideal: Mill seems to hold out before us the vision of a person who aims at, and achieves, true beliefs held with an understanding both of their content and of the grounds on which they rest. If that ideal can be questioned, by juxtaposing it with other things we are inclined to value and revealing tensions, then there will at least be room for debating Mill's defense of freedom of expression.

When Mill's arguments are transferred directly to the context of scientific research it does appear that there are various ways of probing the ideal: Is it realistic to suppose that inquirers today must continually confront the discarded

2. Wilson, "Academic Vigilantism and the Political Significance of Sociobiology," reprinted in *The Sociobiology Debate,* ed. Caplan; see p. 302.

3. See Stephen Jay Gould, "Biological Potentiality vs. Biological Determinism," in *Ever Since Darwin* (New York: Norton, 1977), 258, and my own *Vaulting Ambition* (Cambridge, Mass.: MIT Press, 1985), 7.

doctrines of the past, rather than forging forward? Is the attainment of scientific truth so significant that it overrides any countervailing considerations from the effects of research on human welfare? The latter question is underscored by the discussions of the past two chapters, for, if the arguments I have offered are correct, the significance of scientific results is entangled with practical concerns, and we cannot appeal to some overarching project whose value transcends all others.

Mill would not have been perturbed by these observations. Although his writings are often viewed as a general defense of free inquiry, the types of opinions under consideration in his arguments are quite special. Behind *On Liberty* stands the long sequence of debates about freedom of religious expression. Mill places those debates on a more general level, taking as his principal topic the opinions that are central to people's articulations of their goals and values, of their main projects and the significance of their lives. Chapter 2 of *On Liberty* follows chapter 1, where Mill makes the foundations of his defense completely explicit: "The only freedom which deserves the name is that of pursuing our own good in our own way, so long as we do not attempt to deprive others of theirs or impede their efforts to obtain it."[4] The importance of free expression and open debate is thus to promote individuals' reflective decisions about the ends of their own lives—so to advance "the permanent interests of man as a progressive being"[5]—and the Millian ideal of the agent who recognizes the grounds of his beliefs, fully understands the content of those beliefs, and has had the opportunity to test those beliefs against rivals, acquires its importance precisely because the beliefs in question are those that structure his projects and aspirations.[6] When Mill is understood in this way, the questions of the last paragraph become irrelevant. But, by the same token, there is no longer a direct argument from the precepts he elaborated and defended to the freedom of scientific inquiry.

In fact, we can go further. To take seriously Mill's point that the freedom to which we aspire is the freedom to define and pursue our own vision of the good is to recognize the possibility that the unconstrained pursuit of inquiry might sometimes interfere with the most important kind of freedom, at least for some members of society. So we can envisage a Millian argument *against* freedom of inquiry, one that proceeds by trying to show that certain types of research would be likely to undermine a more fundamental freedom. I aim to articulate this argument, to expose its force and its limits.

4. Mill, *On Liberty* (Indianapolis: Hackett, 1992), 16–17.

5. Mill, *On Liberty*, 14.

6. Here I am in agreement both with Alan Ryan's insightful essay "Mill in a Liberal Landscape," in *The Cambridge Companion to Mill*, ed. J. Skorupski (Cambridge: Cambridge University Press, 1998) (see especially pp. 507, 509, 510), and with Isaiah Berlin's "John Stuart Mill and the Ends of Life," in *Four Essays on Liberty* (Oxford: Oxford University Press, 1969).

Concerns about the social impact of research can be developed in at least three different ways. The strongest, and most ambitious, version of the argument proceeds directly from the difference between the ideal of formulating and pursuing one's own plan and the goals of scientific inquiry, without any further epistemological assumptions. Thus it might be suggested that, were we to recognize certain kinds of truths, the impact on some people would be to erode their sense of worth and to make it difficult, even impossible, for them to frame a conception of their lives as valuable. For the moment, I shall set such considerations on one side; they will occupy us later in a more general context (chapters 12–13).

Alternatively, instead of supposing inquiry will (eventually) deliver the truth, we can take a more realistic (less rosy) view of our epistemic prospects than Mill, his predecessors, and many of his successors are inclined to do. When the expression of unpopular doctrines is defended on the grounds that the clash of views is healthy, it often seems that the defenders take for granted that "truth will out," at least in the long run. Recognizing that research is fallible, as well as socially consequential, we may start to elaborate a critique of some lines of inquiry.

In its most minimal form, the critique need not challenge the value of free inquiry. Those who replied to Wilson's defense of human sociobiology often pointed out that they were concerned with the evidence for the controversial conclusions, and that political considerations were relevant precisely because when the potential consequences are grave, standards of evidence must go up. I'll now try to show that the sociobiology debate offers an opportunity for developing a more ambitious line of argument.

Suppose we envisage scientific investigations as taking place within a society in which there are significant inequalities with respect to well-being. Members of a particular group within this society, a group I'll refer to as "the underprivileged," have lives that go substantially less well than is typical in the rest of the society. This relative assessment of the quality of their lives may turn on obvious economic disadvantages, lower life expectancy, or restricted access to coveted opportunities and positions. Moreover, the reduced average quality of life for the underprivileged is partially caused by the fact that, in the past, it was widely believed that those with characteristics prevalent within the group were naturally inferior or that such people were only fitted for a narrow range of opportunities and positions. Residual forms of this belief are still present, although the belief is repudiated in most public discourse.

Imagine further that some scientific investigations conducted within the society might be taken to support conclusions that bear on the officially discarded belief. Specifically, let the belief in question be, "People with a particular characteristic (call it *C*) are naturally less well-suited to a particular role (call it *R*)," and suppose that an area of science *S* might yield evidence for or against this view. The impact of pursuing *S* and uncovering the evidence is *politically asymmetrical*, in that both the following conditions obtain:

(a) If the evidence is taken to favor the hypothesis that those with *C* are naturally less well-suited to *R*, then there will be a change in the general attitudes of members of the society toward those with *C*, constituting (at least) a partial reversion to the old state of belief; if the evidence is taken to favor the negation of this hypothesis, there will be no significant further eradication of the residues of the old belief.

(b) If the belief that those with *C* are naturally less well-suited to *R* again becomes widespread, then the quality of the lives of those with *C*—the underprivileged—will be further reduced, partly through the withdrawal of existing programs of social aid, partly through the public expression of the attitude that those with *C* are inferior to those who lack *C*; unless there is significant further eradication of the residues of the old belief, there will be no notable improvement in the lot of the underprivileged from pursuit of *S*.

Recognition of the political asymmetry lies behind the modest argument, outlined above, according to which standards of evidence must go up when the consequences of being wrong are more serious.

However, assume also that the society's pursuit of *S* will be *epistemically asymmetrical*, in that people will always take the belief to have more support than it deserves. More precisely:

(c) There will be significant differences between the probabilities assigned to the hypothesis that people with *C* are less well-suited to *R* and the probabilities that would be assigned by using the most reliable methods for assessing evidence; the probabilities assigned to the hypothesis by members of the society will typically exceed the probabilities that reliable methods would yield, and the probabilities assigned to the negation of the hypothesis will be correspondingly deflated.

Although there are already hints of danger for the underprivileged, troublesome consequences aren't inevitable. Evidentiary matters about the effects of having *C* might be clear-cut, favoring the egalitarian conclusion to a large enough extent to outweigh the bias towards the hypothesis.

Suppose, however, this isn't so. If the issues surrounding the impact of having *C* are confusing or complicated, and if the bias towards overestimating the support for an antiegalitarian answer is sufficiently strong, then the underprivileged are indeed threatened by the pursuit of *S*. Specifically, assume that

(d) With high probability, the evidence obtained from pursuit of *S* will be indecisive, in that the most reliable methods of assessing that evidence would assign a probability of roughly 0.5 to the hypothesis.

(e) The bias in favor of the hypothesis is so strong that most members of the society will take evidence that, when assessed by the most reliable methods, would yield a probability for the hypothesis of roughly 0.5 to confer a probability close to 1 on the hypothesis.

If all these conditions are met, there's a significant probability that the antiegalitarian hypothesis will be taken to be extremely well supported, even though the evidence leaves the issue open, with consequent harm to the underprivileged. There is no chance of any genuine benefit for the underprivileged. From the perspective of the underprivileged, the expected utility of pursuing *S* is thus clearly negative. If we shouldn't engage in ventures that can be expected to decrease the well-being of those who are already worse off than other members of society, we should therefore refrain from engaging in *S*.

This argument is abstract and general. Its burden is that when a certain constellation of conditions is satisfied—the conditions (a)–(e)—the pertinent inquiries ought not to be pursued. I strongly suspect that there are cases in which the conditions obtain, and, indeed, that some of the disputes about human sociobiology and human behavioral genetics satisfy the conditions. If we were to take the underprivileged to be the set of women, the characteristics to consist of biological traits uncontroversially possessed by women and not by men, and the role *R* to be any of a number of prominent, and sought after, positions in American or European society, we could generate plausible instances. Even more obviously the assumptions appear to apply to members of various minority groups—African-Americans in the United States, West Indians in Britain, immigrants from North African and Near Eastern countries in European nations.

Consider, first, the political asymmetry. What would be the likely impact of widespread acceptance of inegalitarian conclusions—say that women, "by their nature," lack the competitive urge or the commitment to career to occupy challenging positions, or that minorities have genetic predispositions to lower intelligence? Surely the most predictable results would be the withdrawal of resources from any current efforts to try to equalize opportunity for filling *R*, and a diminution of self-respect and of motivation among the underprivileged. It is hardly cynical to believe that the supposedly scientific findings would inspire policymakers to "stop trying to do the impossible"—instead of "rubbing against the grain of human nature" they would save money now spent on wasteful public programs. Nor is it unreasonable to think that the psychological effects on members of the underprivileged would be damaging, either because they accede to the conclusion that they are less worthy than other members of society, or because they view this as a common perception of their status and thus develop a sense of alienation. At best, these deleterious consequences would be offset by an allocation of public funds to respond to what would now be regarded as the *real* needs and potential of the underprivileged—although it's not entirely obvious what programs of this type would do. Not only are the hypothetical gains extremely nebulous, but it's also far from clear that contemporary affluent societies have much political will for this type of expenditure.

Recent debates about inegalitarian claims support other aspects of the political asymmetry. When evidence is announced in favor of equality, the effect is only to offset whatever damage has been done by more flamboyant presentations of the case for inequality. Defenses of "natural inequalities" typically outsell the egalitarian competition. Furthermore, when rejoinders are published there is no groundswell of enthusiasm in favor of investing more resources in attempts to equalize social roles.

These remarks amount only to a prima facie case. A lot of detailed sociological work would be needed to show that (a) and (b) are satisfied in the scientific controversies about sex and race. Hence it would be reasonable for a defender of research into "the biological bases of social inequality" to protest the application of the argument *if that person were prepared to take on the burden of demonstrating that the consequences I have alleged do not ensue.* That is not, of course, how the defense usually goes, and, in what follows my chief aim will be to consider complaints that the general form of argument by appeal to political and epistemic asymmetries is invalid because it overlooks important aspects of inquiry.

With respect to the epistemic asymmetry it's possible to be more definite about the applicability of the argument, for here a wealth of historical studies hammers home the same moral. First, there is ample evidence of a bias in inegalitarian conclusions: patterns clearly discernible in the history of measuring those traits associated with cognitive performance, from the nineteenth century to the present, from the craniometers to the high priests of heritability, display one version of inegalitarianism (typically seen as preposterous by later generations) widely accepted until painstaking work exposes its lack of evidential support, followed by an interval of agnosticism until the next variation makes its appearance. Second, uncovering the flawed inferences underlying claims of a scientific basis for uncomfortable conclusions typically reveals just how complex are the issues with which investigators are trying to wrestle: analytical study of the methods of trying to show genetic differences in intelligence brings out what would be required to support responsible conclusions; examination of ventures in human sociobiology exposes how hard it would be to do it properly. Reliable knowledge about the topics is hard to come by. Combining this observation with the pattern that emerges from the history, the obvious explanation is that, in an epistemically cloudy situation, the probabilities assigned to the inegalitarian hypotheses are inflated, so that sincere investigators incorrectly believe themselves to have found a scientific basis for socially acceptable conclusions. So I think there's good evidence for the pertinent versions of (c), (d), and (e).

I turn now to some obvious criticisms. First comes the worry that the argument I've presented is myopic. Perhaps in focusing on a particular situation, we fail to understand the more general import of defending free inquiry. Recall Wilson's defense of free inquiry in terms of promoting the intellectual health of the community. He can easily be interpreted as warning of the dangerous effects that blocking sensitive investigations might have on a more general policy whose overall consequences are beneficial. So we might indict the argument for failing to recognize the disutility of closing down particular inquiries, a disutility that results from undermining a society-wide practice of fostering free discussion. Our choices ought to have been framed (so the accusation goes) in terms of a social context for scientific research that is thoroughly committed to a policy of free inquiry, and which occasionally encounters the unfortunate consequences my arguments expose, and a social context for research that hampers the freedom of inquiry, that avoids some local unfortunate consequences, but also suppresses valuable inquiries with appreciable losses in utility.

The obvious answer to this challenge is to deny that our choice is between these two contexts. The objection proposes to evade the argument by mimicking a familiar strategy: faced with the fact that breaking a promise might sometimes maximize expected utility, moral philosophers sometimes suggest that the *rule* of keeping promises promotes well-being and that breaking a promise on a particular occasion would undermine the rule. Unfortunately, the suggestion faces an obvious reply: why should we not adopt a practice of promise-keeping except in situations where it's clear that breaking a promise would maximize expected utility? In similar fashion, the scientific community might be committed to a practice of free inquiry except in situations in which it's clear that certain investigations will be socially disadvantageous (or disadvantageous for those who are underprivileged).

If it were genuinely difficult to distinguish situations in which pursuing some lines of inquiry could be expected to be socially damaging, there might be reason to think that a policy of admitting limits on inquiry would quickly decay to the detriment of society's intellectual health. We begin with good intentions to bar certain investigations but, in allowing the social consequences of an inquiry to determine its legitimacy, we enter a zone in which it's easy to lose our way, ultimately retreating from lines of research that would have proved valuable. Yet the arguments of the form we're considering plainly allow for definite instances, cases in which it's possible to judge that the expected utility of the pursuit of an inquiry is negative (or negative for those who are worst off), and we could block the alleged slide by adopting a policy of only abandoning inquiries when it's clear that the social consequences of pursuit are deleterious (free inquiry would be given the benefit of any doubt). The objection is right to remind us of the broader context in which decisions about the value of free inquiry should be made, but, so far at least, it seems possible to accommodate the point while allowing that some instantiations of the argument are cogent.

Consider a second objection, one that tries to subvert the argument by recalling the historical sources of the beliefs it employs. At any number of stages in the history of the sciences, people with values that were threatened by a particular line of investigation could have contended that the inquiry in question was likely to bring nothing but loss. Imagine committed Aristotelians campaigning against further efforts to determine the earth's motions, or devout Victorians objecting to "speculations" on the origins of species. Had the argument I've given been influential at earlier stages of inquiry, we would have forfeited enormous epistemic advantages. Precisely because we have liberated ourselves from the ideas of our predecessors, through allowing inquiry to undermine accepted beliefs, we are now in a position to make the kinds of evaluations on which the argument depends. Our values have themselves been shaped by the overthrow of previous systems of belief, systems that would have accepted the inequalities in contemporary society with equanimity. Consider, for example, the version of the argument that attacks research into racial differences in intelligence. The recognition that there would be costs if people classified as belonging to minority races were told that authoritative science had established their intellectual inferiority itself depends on a process through which people with particular superficial features and of particular descent were recognized fully as people, a process that depended on the possibility of free inquiry into unpopular topics.

Although this line of reasoning appears plausible, it rests on a number of controversial assumptions. The final step can be debated by questioning the role the sciences have actually played in fostering the acceptance of disadvantaged minorities. The chief defect of the objection lies, however, in the similarity it suggests between the heroic scientific liberators of the past and those who would investigate natural inequalities in the present. People who publish findings purporting to show that behavioral differences stem from matters of race or sex often portray themselves as opposing widely held views in the interest of truth. But do Galileo's would-be successors don his mantle legitimately?

Of course, what matters is *significant* truth, and there are serious issues about why the favored lines of inquiry should count as significant. At this stage, however, I want to focus on a different presupposition of the attempted defense. In understanding the epistemic asymmetry, we recognize a bias towards accepting inegalitarian conclusions because they resonate with attitudes publicly denied but nonetheless present in contemporary societies. Many champions of unpopular inquiries correctly believe their conclusions oppose doctrines affirmed by their colleagues (perhaps even by almost all of those working in the areas related to their discussions) and often upheld by the parts of the media with the strongest intellectual credentials. Their defenses typically fail to mention that there is a broad tendency to believe quite contrary things in private, that the views defended conform to inclinations that voters and public officials harbor and that may even be espoused by those who profess quite different views. In consequence, there's a deep disanalogy between contemporary investigators of racial difference (say) and the scientists of the past who defied the orthodoxies of their age.

Let's say that a belief is part of a *total* consensus just in case almost everyone in the pertinent society accepts it (or is prepared to defer to people who accept it), that a belief is part of an *official* consensus if it is publicly professed by everyone (or if people are at least prepared to defer publicly to people who publicly profess it), that it is part of an *academic* consensus if it is held by almost everyone within the academic community, and that it is part of a *lay* consensus if it is held by almost everyone outside the academic community. Galileo and Darwin opposed total consensus in their communities, and there were powerful biases *against* adopting their conclusions; thus the conditions for applying the argument I've reconstructed to them are not satisfied, and the social disutility of their inquiries can no longer be calculated in the same fashion. Contemporary investigators who claim important differences due to race or sex surely oppose an official academic consensus, and perhaps are at odds with both official consensus and academic consensus. It would be too strong to claim that there is a lay consensus on an inegalitarian conclusion inconsistent with the official academic consensus, but, outside the academy, there are sufficiently powerful inclinations to accept inegalitarian beliefs, held by sufficiently powerful people, to suggest that there will be an epistemic bias in favor of the inegalitarian conclusions, and that these conclusions are likely to be implemented in social policies. Furthermore, there may well be scientists whose embrace of egalitarian claims is sufficiently shallow that they too will be disposed to take indecisive evidence as demonstrating important differences.

Scientists quite understandably bridle at the thought that their research will have to conform to standards of "political correctness," so it's important to understand the exact nature of the argument. Recognizing that some types of research bear on struggles to achieve social justice, *and that there is a schizophrenic moral consciousness in which public "politically correct" attitudes coexist with inclinations to quite opposite beliefs*, we should see the impact of the research as affected by both a political asymmetry and an epistemic asymmetry. Instead of lumping together quite disparate examples from the history of science, it's important to focus on the special conditions the argument discerns in our contemporary predicament. The Millian arena, in which conflicting ideas battle for public approval on epistemically equal terms, and in which the bystanders are never hurt by the nature of the conflict, is a splendid ideal, but it would be quite naïve to think that all pieces of controversial research are discussed in anything like this ideal arena.

The last objection I'll consider may be the most obvious. Perhaps all that the argument shows is the error of a consequentialist treatment of these questions—we go astray in thinking that decisions about the merits of inquiry can be judged by attending to the expected consequences. Of course, the main versions I've considered already incorporate the most prominent concerns about utilitarianism, in that they base judgments on the expected utility for the least fortunate (the underprivileged). Ironically, consequentialism is most sympathetic to inquiry into socially charged topics when we *ignore* the objections. If the response is to succeed, it must propose there's a moral basis for pursuing investigations independently of the impact on the underprivileged. One way to develop that idea is to suppose we have a duty to try to ascertain significant truths about nature. Can this duty override worries about the consequences for the unfortunate?

I think not. Far less controversial than any duty to seek the truth is the duty to care for those whose lives already go less well and to protect them against foreseeable occurrences that would further decrease their well-being. We should recognize a clash of duties whose relative importance must be assessed. To oppose the argument, one must believe that the duty to seek the truth is so strong that it is binding, even in situations that will adversely affect the underprivileged, that will offer little prospect for gaining knowledge, and that will afford considerable opportunity for error.

A different way of opposing the consequentialist framework would be to insist that the project of improving the well-being of the disadvantaged can't be allowed to interfere with rights to free inquiry. This libertarian response would abandon both the consequentialism of the argument and the attempt I've made to avoid typical foibles of consequentialism by focusing on the well-being of the least well-off. Any libertarian defense would thus have to claim that the distribution of rights doesn't matter, that if, through historical contingencies, subgroups of the population have been deprived of various rights we can't seek to remedy the situation by abridging the rights others enjoy, even if doing so would limit rights in small ways to enhance dramatically the ability of the disadvantaged to exercise rights others take for granted. It would also have to argue that the right to free inquiry is fundamental, that it overrides important rights of those who suffer from the pursuit of inquiries that reinforce incorrect stereotypes. I think it doubtful either of these challenges (let alone both) can be met, but, in any event, there is a simpler antilibertarian argument. Respecting rights comes at a price, and it's important that the price be distributed fairly. In situations where free inquiry would unfairly increase the burden on those who are already disadvantaged, there can be no right to free inquiry.

If one seeks to reject the argument by abandoning its utilitarian framework, the best approach seems not to be to invoke implausibly strong collective duties or uninfringeable rights, but to suggest instead that freedom of expression is required for the deepest and most important kind of human well-being, to return, in effect, to Mill's own conception of "the permanent interests of man as a progressive being." Couched in Mill's own terms, where the focus is on our capacity for choosing our own vision of the good and for planning how to achieve it, this is quite promising. One might argue that free inquiry is needed if we are to discover what is best or most worthwhile and how to create the conditions most conducive to its realization.[7] As I hinted earlier, these considerations do support the ideal of freedom of inquiry *to the extent that it promotes human reflection and deliberation.* But they do not provide any escape from the argument about free scientific inquiry.

The difficulty on which earlier attempts at evasion have foundered is the conflict between a relatively abstract value (the attainment of truth) and the concrete ways in which some people's lives are diminished by the purveying of inegalitarian conclusions. It would be easy to conceive the latest version in the same terms, taking it to oppose the fundamental interests that ground freedom of expression to the demands of equality, and so heading for a familiar stalemate. But this would be mistaken. The issue isn't how we weigh competing fundamental values (freedom vs. equality) but rather how we require whatever values are seen as most fundamental to be distributed. Champions of free inquiry often view it as a precondition of human well-being because they think in terms of an abstract human subject whose deliberative capacities are enhanced by open discussion—all is calm, serene and unthreatening. How representative is this abstract subject of the actual people whose deliberations would be affected by the actual pursuit of the forms of inquiry about which there is dispute? Once this question is posed, we begin to understand that the structure of the argument in its consequentialist form can be replicated precisely because the *absence* of particular kinds of inquiries would enhance the deliberative capacities of those for whom deliberation is currently most constrained. We can agree with Mill and his successors that the freedom to deliberate is fundamental, and, in consequence, *adopt just the argument I have given on the grounds that it promotes a fair distribution of this fundamental freedom.* To make this more concrete, we can compare the controversial inquiries to other instances in which the value of free expression is undermined by the importance of securing the deliberative freedom of the disadvantaged. Consider hate speech. For certain kinds of verbal performances it's at least arguable that the effect isn't to broaden the possibili-

7. Variations of this general approach are offered by Joshua Cohen, "Freedom of Expression," *Philosophy and Public Affairs*, 22, 1993, 207–263, and David Brink, "Millian Principles, Freedom of Expression, and Hate Speech," *Legal Theory*, 7, 2001, 119–57.

ties of dialogue but rather to exclude the victims of the speech from discussion, thus cramping their deliberative opportunities. Obscene racial epithets don't invite a calm rejoinder that would open up new intellectual vistas for both aggressor and victim; they are intended to banish some people from public fora, and they often succeed in doing so.

Mill's own principles thus support the conclusion that certain forms of inquiry ought not to be pursued, undercutting the popular—and, I believe, well-intentioned—view that the free pursuit of inquiry is always a good thing. Yet, as Mill saw as clearly as anyone, the fact that we ought not to pursue a particular course of action doesn't mean that there should be a publicly enforceable ban. We can thus distinguish two potential conclusions of the argument I've offered, one which supposes that certain types of research should not be undertaken and another that takes the further step of claiming these inquiries should be proscribed. The distinction is important. For the argument that, when conditions (a)–(e) are satisfied, there are moral grounds for refraining from inquiry is cogent. Demanding a ban on inquiry under such conditions would be to take a further, illegitimate, step.

Mill proposed that the scope of law is limited to those instances in which the prohibited action would cause harm to others, and one might initially think that this proposal clears the way for a ban on some types of inquiry, namely those likely to erode further the status of the underprivileged. The problem with a ban does not stem from the Millian proposal, but from the consequences of instituting it: in short, the "cure" is worse than the "disease." More exactly, the very conditions that underlie the asymmetries on which the argument draws ensure that officially restricting free inquiry would exacerbate the social problems.

In a world where (for example) research into race differences in I.Q. is banned, the residues of belief in the inferiority of the members of certain races are reinforced by the idea that official ideology has stepped in to conceal an uncomfortable truth. Prejudice can be buttressed as those who oppose the ban proclaim themselves to be gallant heirs of Galileo. When the Caucasian child asks why research into differences between racial groups is not allowed, a superficially plausible answer will be that everyone knows what the research would show and that people are unwilling to face the unpleasant truth. Proscribing the research has consequences of the same general kind as allowing it—except that they are probably worse. *So long as the epistemic asymmetry is not clearly appreciated*, champions of the research will always ask (rhetorically), "If there is genuine equality, why not try to demonstrate it?" From the perspective I've been defending, there's an answer to the question, and it lies in pointing out conditions (c), (d), and (e). If the answer were widely accepted, there would be no need for a ban; when the answer is not widely accepted, any ban would be seen as illegitimate.

I doubt that more limited solutions, such as the withdrawal of public funds from support of certain kinds of projects, will fare any better. So long as the conditions driving the argument are not appreciated, champions of the forms of inquiry that should be eschewed can always make use of the rhetoric of freedom to portray themselves as victims of an illegitimate public policy of stifling the truth. The consequences of *any* type of official intervention are thus likely to be counterproductive—and this may even extend to presentations (like this one) of the harms inquiry may engender.

These gloomy reflections do not touch the argument that the research under scrutiny is morally unjustified, although they leave us with a dilemma about what to do once we recognize the point. One obvious suggestion is that we view the conclusion simply as a moral imperative. Responsible scientists ought to ponder the consequences of their own work and refrain from research into areas to which the conditions of the argument apply. Yet the very considerations that inspire the argument make it apparent that we can't hope that scientific self-scrutiny will be efficacious. The pressures that build the epistemic asymmetry, most prominently the temptation to gain a large audience and to influence public opinion by defending "unpopular" views, make it highly likely that scientists fascinated by the thought of exploring differences due to sex, gender, or race will read the evidence very differently: some will deny that the history of research into these areas has revealed a pattern of shoddy work, widely accepted as supporting inegalitarian conclusions until critics expose the deficiencies, seeing instead the tragedy of bold predecessors pilloried for peccadilloes by a politically biased establishment; others may admit the troubles of the past, insisting that things are different *this time*, that the brave new method will finally deliver the goods. I believe a sober review of the history of research into racial and sexual differences supports the view recorded in the argument, and thus any attempts to read that history differently embody just that epistemic bias that the argument diagnoses. Moreover, the sense that investigators now have new tools for conducting their inquiries should be coupled with a clear understanding that others have been similarly optimistic in the past, and that epistemic bias may lead one to overrate the force of the "latest findings." Nonetheless, from the perspective of the argument, it's to be expected that those attracted to research in these areas will find ways of denying key premises, so their attempts at moral reflection, however sincere, will not persuade them that they should abstain.

If this is correct, we can neither use the argument to support public discouragement of certain types of inquiry, nor expect that private moral reflection will make the problem go away. So what implications, if any, does the argument have? What's to be done?

To repeat: the Millian conception of the arena in which doctrines compete equally and in which public expression of those doctrines causes no harm is a splendid ideal. Unfortunately, it sometimes inspires people to take a naively optimistic view about the actual pursuit of inquiry. To accept the argument I have offered should *not* be to conclude that there ought to be public censorship of inquiry. Instead, it ought to provoke us to pursue questions about the social surroundings of our investigations: What are the conditions under which the Millian arena functions properly? What kinds of social factors cause those conditions to break down? What can be done to prevent the breakdown?

There are two polar views, both of which ought to be resisted. One claims the Millian arena always functions properly, yielding eventually secure knowledge whose value outweighs whatever harm has been caused in the fray. The other denies that the Millian arena ever works in the way it is supposed to, contending that what we take for a competition among ideas on their evidential merits is always a thoroughly political struggle. The reasoning of this chapter is intended to show what is wrong with the first view. My quarrel with the second begins with the modest realism espoused in chapters 2 and 3. The position I favor is that we sometimes achieve true beliefs about nature, that, when we do so, we often proceed by reliable means and gain knowledge, and that, in some of these instances, lively debate between partisans of different positions is instrumental in our attaining knowledge—in much the way Mill envisaged. Unfortunately, this is not always possible, and, as the argument recognizes, we're sometimes victims of epistemic biases. But our plight isn't hopeless and we may be able to identify those epistemic biases, thus avoiding those clashes of ideas that would be genuinely harmful. The failures of the Millian arena are local. With respect to some issues, open debate will generate opinions with the virtues Mill so lucidly characterized, opinions that are closer to the truth, held with understanding on the basis of reliable methods that are clearly recognized. Using the history of our inquiries as a guide, we can come to distinguish these instances from others in which the arena fails to function, and, perhaps, on the basis of the distinction, we can improve our epistemic condition.

We have reason to be confident in the claims we make in some areas of inquiry, but there's no guarantee that the methods that spawned those claims will apply generally to yield knowledge in all the fields we'd like to investigate. Once this point is appreciated, we obtain just the perspective on the Millian arena I wish to commend. Instead of believing that the "contest of ideas" will always guide us to the truth, we can see that, in some instances, the contest might indeed be helpful, while in others it will not—whether because the problems are too hard or we suffer from biases we haven't yet learned to eradicate (and which may be ineradicable). Our best strategy is not to start from the assumption that free inquiry will always be a good thing, but rather to use the style of argument I've developed, in tandem with serious analysis of the successes and failures of our past ventures to try to improve our methods of inquiry (and to abstain from those investigations that can be predicted to bring only trouble).

The passages from Wilson with which I began make it apparent that this is not a standard way of considering scientific research. We aren't used to thinking about the ways in which our attempts to achieve knowledge, and the track records of their successes and failures, impinge on people's values and interests. I suggest that this is because of a tension between the science that is practiced in democratic societies and the underlying ideals of those societies. I formulate this as the thesis that *science is not well-ordered.* The task of the next three chapters is to understand what well-ordered science would be.

NINE

Organizing Inquiry

HOW SHOULD INQUIRY BE ORGANIZED so as to fulfill its proper function? At the dawn of modern science, both Bacon and Descartes attempted to address the question. They saw the history of attempts to understand the natural world as dominated for two millennia by the faulty doctrines and methods of Aristotelianism, and resolved that such stagnation should never occur again. Offering a common diagnosis, both saw the need for a method of discovery future inquirers could follow, whatever questions they were concerned to resolve, and a method of justification that would specify exactly when answers should be accepted; in focusing on the latter project, they hoped to forestall the premature adoption of views that would distort future research. Of course, their proposals for answering questions about discovery and justification differed. Descartes emphasized the virtues of rational analysis, both to direct the mind in discovering solutions to problems and to frame the options experiments might discriminate. By contrast, Bacon stressed the importance of unprejudiced observation and the patient accumulation of empirical data.

Neither of these proposals wins contemporary acceptance, although both have left their mark. Yet Descartes and Bacon continue to shape—and to limit—contemporary discussions of "proper" inquiry. From the seventeenth century to the present, their main questions—What is the right method of discovery? What is the logic of justification?—have dominated reflections about the sciences. In the twentieth century, the first question fell out of favor as an influential group of philosophers argued that there was no general method of scientific discovery. Recently, emphasis on the serendipity of discovery has given way to a more refined appreciation of the methodical ways in which investigations typically proceed, and scholars have turned to new formal techniques to resurrect the concept of a method of discovery.

For the past seventy years, however, the central normative question about science has concerned the logic of justification. How can we identify the conditions under which statements—particularly universal generalizations and statements about entities remote from observation—are properly accepted? The dominance of this question suggests a way of thinking about inquiry. To a first approximation, philosophers of science have written as if inquiry would fulfill its proper function provided all those who engage in it live up to the standards set by an ideal methodology, accepting only those statements that are genuinely justified in light of a correct logic of justification.

It is remarkable just how much this conception leaves out. Besides issues about methods of discovery (now regaining some respectability), there are other obvious omissions. We typically evaluate an enterprise by considering how well it achieves, or can be expected to achieve, its aims. As some philosophers have seen, an assessment of the sciences that ignores the fact that we risk error in hopes of gaining information cannot be right. Unless the logic of justification goes beyond the popular project of trying to specify the conditions under which particular types of conclusions are likely to be true, it will fail to see that what matters in well-organized inquiry is the standard of gaining significant truth. Even when discussions of the sciences appreciate the elementary point that the most reliable means of ensuring that the statements one accepts are not false is to accept nothing whatsoever, the resultant approach to issues of justification frequently supposes there is some context-independent good thing for which it's worth risking error. Philosophers conjure up general measures of content, or explanatory power, or information, and then puzzle about how, in general, we should weigh the chance of being mistaken against the abstract benefits they favor. But what inquiry seeks is significant truth, and, as we've seen, significance is thoroughly context-dependent. There is, then, no general problem of trading significance against chances of truth, and, accordingly, no general solution. The right strategy is to frame the problem, at the start, in terms of how well-functioning inquiry promotes the acquisition of significant truth, explicitly acknowledging the variability of standards of significance.

Even more obviously limiting is the traditional focus on individuals. Unlike Descartes, Bacon already recognized that inquiry is a collective affair, and his ideas about the social character of proper inquiry were influential in the formation of the Royal Society. (Some of those ideas will occupy us briefly in chapter 11.) The community of inquirers cannot simply be viewed as a magnified version of the individual. Indeed, when we appreciate how changes in the sciences stem from the combined efforts of individuals, we may be led to adjust our views concerning how single investigators should behave. Suppose methodology in the individualistic tradition has succeeded in picking out the right rules for accepting hypotheses (or assigning probabilities to hypotheses) on the basis of evidence. A community of scientists, each of whom follows these rules and all of whom have access to exactly the same evidence will be homogeneous in its opinions. Yet, possibly for Millian reasons (canvassed in the last chapter), one might doubt if homogeneity is the best epistemic policy. Maybe the collective attainment of truth (or significant truth) would be advanced if some members of the community were to disagree. So an account of what individuals should do won't automatically tell us when collective inquiry is working properly.

The discussion of the last chapter identified a third major omission. Suppose we broadened the traditional perspective of concentrating on the attainment of truth by individual inquirers, considering the collective pursuit of significant truth (where significance is understood in the way suggested in chapter 6). Even that would not take into account the possibility that the search for significance can conflict with other important values. To assess the proper functioning of scientific inquiry we must consider if collective research is organized in a way to promote our collective values in the most encompassing sense. To enclose scientific research so that the standard to which it is held is the collective acquisition of epistemically significant truth is to acquiesce in the myth of purity, and, as the last chapter tried to show, that approach will allow projects that are morally suspect.

Traditional philosophy of science has offered welcome clarifications of important concepts and principles, and we should value its insights. But, as I've complained, the dominant perspective has provided a very narrow normative perspective. Perhaps that is why the classical view of science has attracted criticism, generating the polar pair of unacceptable images described in chapter 1. What follows is an attempt to put back some of the considerations that have been slighted.

When we lived in California, Bertie, the much-loved family dog, would come and go freely between the house and the fenced backyard. One afternoon, two or three hours before sunset, when coyotes become active, someone came to read the meter and left the gate to the yard open. When Bertie went out, he spotted the chance of adventure and set off to explore. Discovering his absence we had to formulate a plan for finding him (quickly, since he's too small to take on coyotes). We normally walked him along one of two routes, which we assumed were the places he'd be most likely to go. One route we estimated to be a more probable path than the other. There were four of us. How to proceed?

The problem, of course, is underdescribed, for I haven't specified the probabilities of finding him along either route, the chances of his doing something different, the number of people needed to scout each route thoroughly, and so forth. But one point ought to be clear. If the four of us had stayed together, even if we went over the most likely route, that would probably have been a very bad strategy. What we'd like to have known, of course, provided it could be identified with negligible costs of time to calculate it, is the strategy that would maximize our chances of finding Bertie within the next two hours. We probably didn't manage that. We did avoid the really bad strategy of staying together, though, and it's quite possible that our approach yielded a probability of success close to the maximum value. (In any event, we found Bertie quite safe, and brought him home before dusk.)

This homely example not only shows clearly how a well-organized inquiry can require investigators to do different things but also suggests a criterion for appraising strategies for inquiry. Given the information available to us, we want our efforts to be organized so as to maximize our chances of attaining our goal. Similar problems arise in scientific contexts, and, at least prima facie, they are associated with a similar criterion of success and sometimes allow explicit analysis. Suppose, for example, that a community of chemists hopes to fathom the structure of a very important molecule (VIM). Two methods are available. One is assessed as highly likely to succeed if pursued with sufficient vigor, and as not necessarily slow. The other is risky: it might deliver the answer quickly or might be quite inadequate to the task. If the community contains enough members, then, just as in the case of the search for Bertie, the best strategy is to divide the labor.

Our dog-finding efforts began with a discussion whose upshot was to designate an agreed-upon role for each of us. We might have reached the same division of effort in a different way: imagine the family returned home sequentially, and each person left a note explaining the situation and the searches in progress so far. Scientists, of course, do not typically assemble to agree on roles for attacking a problem (although this type of explicit cooperation is not unprecedented, especially in conditions of war). But it is not hard to see how a community might mimic the second way of distributing the effort among its members: as newcomers enter the field they take stock of the ways in which central problems are currently being pursued and adjust their own research to maximize the chance that the community will succeed. Yet this requires a type of high-mindedness that may be rare. Can we expect inquirers to be devoted to the common project of advancing knowledge in the same way the family is dedicated to reclaiming its lost dog? That seems implausible. Scientists do not often declare in public that they intend to pursue some unpromising line of inquiry because their doing so will advance the community project of solving a problem—and I doubt if private commitments of this kind are more common. But this may not matter.

For there are other ways in which a community can reach a satisfactory distribution of its research efforts. Suppose everybody in the group of chemists recognizes that whoever fathoms the structure of VIM will receive great kudos, maybe even win a much-coveted prize. For each individual scientist the desire to be the one to solve the problem (and perhaps to win the prize) is the dominant motivation. Now we can imagine some of the chemists surveying the current distribution of effort, and, believing that there is too much competition among the followers of one method, switching to the other. The latter may be more risky, but, because of the lower competition, the scientist's chance of winning the race for the solution would go up. So the community can avoid the disastrous state of homogeneity and even come close to the optimum.

The most important moral of my story so far is the clear delineation of a possibility: we can have a community well-designed for the attainment of epistemic goals in which social institutions we might have viewed as irrelevant, even contrary, to those goals (attributions of credit, prizes) are tailored to motivations typically viewed as antithetical to the goals (desire for personal glory). We might hope to go further. If we could identify a recurrent set of scientific predicaments, then we might be able to embed this particular analysis in a broader study, one that would show the effects of various types of social arrangements, given prevalent human motivations, and pick out the package that would best promote the community's attainment of truth. Finally, we could hope to expand the set of goals beyond the epistemic, repeating the same style of analysis with respect to this broader conception of value. So we'd arrive at an explicit account of the proper functioning of inquiry.

This is far too optimistic. The examples I've offered were tractable because it was possible to specify a *local* epistemic goal. Our family wanted to find the dog before dusk, the chemists wanted to reach a state in which one of them recognized the structure of VIM. Trouble sets in when we try to think more globally.

If there's no context-independent notion of significance, then any attempt to develop a measure of epistemic value—the quantity inquiry is to be designed to maximize—will embody current ideas about epistemic significance. As the notion of epistemic significance evolves between the time at which the possible achievements are assessed and the time at which new knowledge is gained—almost certainly with important surprises—the measure assigned retrospectively may differ quite radically from that proposed in advance. Perhaps however we tried to organize inquiry we'd be doomed to regret our choices. Or maybe almost any decisions would generate future states in which we were happy with what had been attained.

Even were we able to offer definite specifications of our epistemic goals, posing and solving the pertinent optimization problem would still be difficult. Focusing on a recurrent decision situation, such as the one in which the community has two available methods for tackling an important problem, might allow us to show that a particular social institution, like the public awarding of credit to the first solver, would lead to a good distribution of effort. But this success might be offset by comparable failures if the circumstances were slightly different, and without detailed knowledge of the likely frequencies with which situations of various kinds arise, there would be no basis for viewing the institution in question as beneficial. Worse still, without analyzing the impact of the institution across the total range of contexts in which it might play a causal role, it's impossible to discern its overall effect. The dangers of local optimization are familiar from evolutionary biology (and other areas of inquiry). Moreover, as the last chapter already suggested, the pressure to earn public acclaim, which might prove valuable in promoting cognitive diversity, can also have much less salutary consequences—for example leading researchers to leap to conclusions that resonate with popular prejudices.

These points underscore the difficulty of elaborating a fully general social methodology. We would like the sciences to be well organized so we could learn as much as possible about the world as efficiently as possible, but it may not be possible to formulate any serious proposals about optimal organization until our knowledge of nature is vastly richer than it is; perhaps in order to identify the recurrent predicaments inquirers face and to estimate reliably the chances of success of various methods, we already need to know most of the things we want our inquiries to disclose. For all that, reflections on social methodology are far from useless, in that they can show us how to avoid really bad strategies and sometimes reveal to us problems that we wouldn't otherwise have seen or benign consequences where we might have harbored suspicions. The best is too much to hope for, but we may aspire to improve our situation.

Consider a sequence of questions. First: what are good policies for individuals to adopt if they want to learn epistemically significant truths? Second: what are good ways for communities to organize their efforts if they want to promote the collective acquisition of epistemically significant truth? Third: what are good ways for communities to organize inquiry if they want to promote their collective values (including, but normally not exhausted by, the acquisition of epistemically significant truth)? The first of these questions is closest to the philosophical tradition of trying to clarify the methods of inquiry, and, because of the context-dependence of the notion of epistemic significance it is already hard to treat both formally and generally. We can look for formal approaches to special instances of it or for a more general approach that aims at a qualitative characterization of promising strategies and pitfalls. Matters are similar, as we have just seen, with respect to the second question. On the face of it, the third question—the issue with which we really ought to be concerned—appears even harder.

For in the first two cases we think we know what we are trying to attain: epistemically significant truth, either for the individual or for the community. When we broaden the perspective to encompass our "collective values," the goal becomes much more nebulous. How does this goal relate to the actual wishes and preferences of the members of a society? How are we to integrate the preferences of different people? Can very different types of value be brought under a single measure? Is it even possible to undertake the local analyses that are available in the more limited epistemic projects? These are all serious concerns, and they will require our attention.

There are two very obvious ways of approaching the problems of the last paragraph. One is to suppose that whatever preferences people actually have, whatever they think about what it would be good for them to pursue, either individually or collectively, some ends are objectively worthy and there are objective relations among these ends. Call this general perspective *objectivism about values* (*objectivism*, for short). Objectivism can concede that there are many different kinds of values, some of them practical, some epistemic, some present, some future. It may even countenance human diversity, supposing some packages of good things are better for some people, different packages for other people. But objectivists think there's a right way of trading the epistemic against the practical—and, more generally, a right way of trading various different types of values for one another—a right way of balancing the present against the future, and a right way of integrating the objective interests of different individuals.

Here's an example (not, I hope, a particularly implausible one) of an objectivist position. It is objectively good for people to develop their talents and to enjoy as much liberty as possible in deciding on and pursuing their chosen goals. With respect to each assignment of various freedoms and resources (food, shelter, education, and so forth) at the different stages of the person's life there is an objective level of value for that person (not necessarily the same for all people). Further, there's an objective way of aggregating the levels of value for each person into a measure of the collective value achieved: let's suppose it consists in adding the individual levels of welfare subject to a function that discounts them if there are pronounced inequalities among individuals. So, with respect to different social arrangements and endeavors, there will be an overall measure of expected value, one that depends objectively on the expected contributions of those arrangements and endeavors to the levels of value achieved by individuals at the successive stages of their lives, that integrates those levels of value across the person's life-span and that aggregates the expectations for the population by adding the individual expectations, subject to the discounting that represents the costs of inequality. Relative to a specified collection of other social arrangements and endeavors, we could now say that a particular way of conducting inquiry within a social context fulfills the proper function of inquiry just in case it furnishes the maximal level of collective expected value attainable within that social context. (Plainly, we'd be happy to settle for ways of conducting inquiry that failed to maximize but that came relatively close.) In other words, inquiry functions well when it can be expected to lead to states in which people have, at the different stages of their lives, the resources and freedoms they need (according to the basic account of what's valuable for the individual), and when the total level of lifelong value, calculated across the society, is high and not marked by large inequalities.

Perhaps others will see how to articulate a conception of this general sort. I don't. My doubts rest on the difficulty of divorcing what is good for a person from that person's own reflective preferences and the kindred problem of ignoring personal preferences in understanding the ways in which different distributions of goods across the stages of a person's life yield overall value. Further, I think that the general problem of understanding how to aggregate individual levels of well-being into a measure of collective welfare, in the ways objectivists propose, is extremely difficult. So I propose something more modest. Individual preferences should form the basis for our understanding of the personal good that inquiry (among other social institutions) is to promote. In moving from the individual to the measurement of value for the society, we should explicitly limit our discussions to societies that honor certain democratic ideals. Hence my approach to the fundamental question, "What is the collective good that inquiry should promote?" will start from a subjectivist view of individual value (using personal preferences as the basis for an account of a person's welfare) and will relate the individual good to the collective good within a framework in which democratic ideals are taken for granted.

The task for the next chapter is to deliver an answer to the fundamental question. I don't pretend that what I offer is the unique best answer—or even that the approach to which I've just committed myself is the preferred way of tackling the question. As I've indicated throughout this chapter, traditional discussions of scientific inquiry adopt a much narrower conception of the standard against which the proper functioning of inquiry should be assessed. The obvious difficulty in attempting to think more broadly about the role of the sciences within society is the lack of any clear conception of what the wider task of inquiry might be. In providing *an* answer I hope to respond to skeptical concerns that there is no coherent wider conception, and also to delineate the area in which a plausible answer may be taken to lie. It would be folly, however, to pretend that I have given convincing reasons for distinguishing my particular proposal from rival members of the family, or that my account of the details is likely to be correct. Others, perhaps, may be able to improve on it.

TEN

Well-Ordered Science

THERE'S A VERY SIMPLE WAY TO DEVELOP the idea that properly functioning inquiry—well-ordered science—should satisfy the preferences of the citizens in the society in which it is practiced. Projects should be pursued just in case they would be favored by a majority vote. Call this "vulgar democracy."

Vulgar democracy doesn't require actual voting. Rather it offers a standard against which we can assess rival schemes for deciding which endeavors are to be undertaken. The idea of calling together the citizenry to cast ballots on each occasion of decision is evidently absurd, but vulgar democracy is only committed to seeking social arrangements (committees of representatives, for example) that we might expect to do well at mimicking the outcomes of the expression of individual preference. Nevertheless, as its name suggests, vulgar democracy is a very bad ideal.

The most obvious deficiency, of course, lies in the fact that people's preferences are often based on impulse or ignorance and thus diverge from favoring what would actually be good for them. Only a moment's reflection is needed to see that the most likely consequence of holding inquiry to the standard of vulgar democracy would be a tyranny of the ignorant, a state in which projects with epistemic significance would often be dismissed, perceptions of short-term benefits would dominate, and resources would be likely to be channeled toward a few "hot topics." Because these consequences plainly diverge from the promotion of collective well-being, vulgar democracy is a bad answer to our question.

How can we do better? I offer a homely analogy. Imagine a family with a free evening and a strong shared wish to spend it together in some form of entertainment. They begin with a number of different proposals, explaining to one another their preferences, the strength of the preferences, the considerations that move them. Each family member learns new things about the character of the various options, and each learns how the others view the possibilities. Nobody wants to do anything that any of the others regards as too unattractive, and they end up with a plan that reflects their collective wishes. Those collective wishes can't be conceived as the wishes that would emerge from a simple vote in the initial state of mutual ignorance; rather, they are the wishes that would be produced by a more intricate negotiation.

In order to use the analogy to articulate an ideal procedure for well-ordered science, we need a clear understanding of the kinds of decisions that will be needed. Let's conceive of ideal inquiry as divided into three phases. At the first phase, decisions are made to commit resources, such as investigators and equipment, in particular amounts to particular projects. The second phase pursues those projects in the most efficient way, subject to moral constraints that rule out certain physically possible options. At the third phase, the results of the various investigations are translated into practical consequences. So there are three different decisions to be made: How are resources initially to be assigned to projects? What are the constraints on morally permissible investigation? How are the results of the investigation to be applied? As we'll see, the first and the third decisions can be approached within a similar framework.

Begin with the first decision. I envisage individuals with different initial preferences coming together, like the family, to discuss the available courses for inquiry to pursue. The first thing to recognize is that, unlike the family, they are likely to begin from a very partial understanding of the possibilities. An obvious remedy for their ignorance is to insist on transmitting information so that each deliberator becomes aware of the significance, epistemic and practical, attaching to potential lines of inquiry. Ideal deliberation must involve presenting the structure of significance graphs, where the multiform sources of significance are revealed. Once this has been accomplished, the deliberators revise their own initial preferences to accommodate the new information. Specifically, I imagine that each considers how possible inquiries might bear on goals that were antecedently adopted. The product of the consideration is a collection of lists of outcomes the deliberators would like scientific inquiry to promote, coupled with some index measuring how intensely they desire those outcomes. Personal preferences have given way to *tutored* personal preferences.

The next step is for ideal deliberators to imitate the imaginary discussion of the family. They exchange their tutored personal preferences, explaining why they want particular outcomes to particular degrees and listening to the explanations given by others. In this process, I assume each is moved by respect for the preferences of others and aims to arrive at a consensus list in which none of the others is substantially underrepresented. The deliberators are committed to seeing the others as having, like themselves, a claim to realize their aspirations, and thus to take seriously the others' descriptions of their preferences and predicaments and the rationales they provide for choosing as they do. Ideal deliberators thus recognize that they are engaged in a long-term sequence of interactions with people whose situations and fundamental wishes may be quite different from their own, and that such people cannot be expected to sacrifice their desires to the preferences of others.

At the end of this exchange, the preferences of each ideal deliberator are again modified, this time to absorb their recognition of the needs of others. The next step is for them to attempt to draw up a list that represents their priorities concerning the outcomes to which inquiry might prove relevant. One possibility is that there is consensus. After coming to understand both the current characteristics of significance graphs and the tutored preferences of other deliberators, each party formulates the same list, assigning exactly the same value to each outcome. If this is so, then the resulting list expresses the collective preferences, and no further accommodation is needed. A second possibility is that some deliberators favor different lists but each is prepared to accept a set of lists as fair and the intersection of the sets is nonempty. Under these circumstances, if the intersection contains a unique member, then that expresses the collective preferences. If not, then the ideal deliberators must decide by vote which of the lists in the intersection is to be preferred. Finally, if the intersection of the sets of lists deliberators accept as fair turns out to be empty, collective preferences are determined by vote on all candidates drawn from the union of these sets of lists.[1]

At this point, our deliberators have formulated the issues they'd like inquiry to address and have indicated the relative weight to be given to these issues. Their formulation, tutored by a clear understanding of the sources of significance for scientific endeavors already completed as well as those that might now be undertaken, can be expected to recognize possibilities for satisfying curiosity as well as opportunities for practical intervention, long-term benefits as well as immediate payoffs. The next step is to assess the possibilities that particular scientific ventures might deliver what the ideal deliberators collectively want. Given a potential line of inquiry that might bear on some items on the collective list, we require an estimate of the chances the desired outcomes will be delivered, and it's appropriate to turn at this point to groups of experts. How are the experts to be identified? I suppose that the ideal deliberators can pick out a

1. It's clear from the extensive literature on social choice theory stemming from Kenneth Arrow's famous impossibility theorem that any procedure like the one described here may lead to counterintuitive conclusions under certain hypothetical circumstances. I don't assume that my suggestion is immune to such problems, but I hope they only arise in sufficiently recherché situations to make the standard I attempt to explicate appropriate for the purposes of characterizing well-ordered science. For lucid discussions of Arrow's theorem and its significance, see Amartya Sen, *Collective Choice and Social Welfare* (San Francisco: Holden Day, 1970).

group of people to whom they defer on scientific matters generally, that this group defers to a particular subgroup with respect to questions in a particular field, that that subgroup defers to a particular sub-subgroup with respect to questions in a particular subfield, and so forth. Further, it's assumed the experts identified are disinterested—or that any members of a group whose personal preferences would be affected by the project under scrutiny are disqualified from participating in the process. If matters are completely straightforward, there'll be consensus (or virtual consensus) at each stage concerning the appropriate people to consult, and these people will agree on exact probabilities with respect to the outcomes of research. In that case, the output of the search for probabilities is just the collection of chances assigned by the groups singled out at the ends of the various chains of deference.

Complications can arise in any of three ways.[2] First, there may be disagreement about to whom deference is warranted. Second, the experts may be unable to do more than assign a range of probabilities, possibly even a wide range. Third, the experts may be divided on which probabilities (or ranges of probabilities) should be assigned. I deal with all these complications in the same general way, namely through being inclusive. If any of them arises, the output of the search for probabilities is no longer a single set of values, but an explicit record of the verdicts offered by different groups, coupled with the extent to which those groups are supported by deliberators who have full information on the current state of inquiry and the past performances of the groups in question. Hence, instead of a simple judgment that the probability a scientific project will yield a particular desired outcome is a certain definite value, we may have a more complex report to the effect that there are several groups of people viewed as experts, that deliberators in full command of the track records of the various groups select particular groups with particular frequencies, that the groups divide in their judgments in specified proportions, and that these judgments assign specified ranges of probability values.[3]

At the next stage, we suppose a disinterested arbitrator uses the information about probabilities just derived, together with the collective wish list, to draw up possible agendas for inquiry. The arbitrator begins by identifying potential levels of investment in inquiry (possibly an infinite number of them). With respect to each level, the task is to pick out either a single assignment of resources to scientific projects best suited to the advancement of the deliberators' collective wishes, given the information about probabilities, or a set of such assignments

2. One might envisage a fourth. Mightn't it turn out that all the people most competent to assess a venture have an interest in the outcome? In the real world, this is an obvious possibility. But, for our purposes, it's enough that there are ideal experts, who share all the knowledge of actual judges but have no personal stakes in the line of inquiry.

3. I am indebted to Stephanie Ruphy for a discussion that brought home to me the potential complexities of the appeal to experts in this context.

representing rival ways of proceeding that cannot be decisively ranked with respect to one another. In the simplest case, when the arbitrator is given point probability values, the decision procedure can be specified precisely: with respect to each budgetary level, one identifies the set of possible distributions of resources among scientific projects compatible with the moral constraints on which the ideal deliberators agree, and picks from this set the option (or set of options) yielding maximal expected utility, where the utilities are generated from the collective wish list and the probabilities obtained from the experts. (Although the process of deciding on moral constraints on inquiry hasn't yet been considered, we'll see below that it's quite independent of the decision to be made by the arbitrator.) If there are disagreements about who the experts are, disagreements among the experts, or latitude in the responsible assignment of probabilities, then the arbitrator must proceed by considering the distributions of resources that would meet both budgetary and moral constraints, subject to different choices for probability values, picking that set of distributions that best fits the views of the majority of those who are most often considered experts.[4]

The last stage of the process consists in a judgment by the ideal deliberators of the appropriate budgetary level and the research agenda to be followed at that budgetary level. Perhaps there is consensus among the ideal deliberators about which level of support for inquiry should be preferred, and perhaps the arbitrator assigns a single distribution of resources among lines of inquiry at that particular level. If that is not so, then the final resolution must be reached by majority vote. The result (whether it comes from consensus or voting) is the course of inquiry that best reflects the wishes of the community the ideal deliberators represent.

As already noted, the procedure just outlined presupposes some agreed-upon constraints on inquiry. I suppose these take a normal form, always stemming from the recognition that a particular way of pursuing inquiry would violate the rights of some individual or group. Any disagreements among the ideal deliberators are thus traceable to different conceptions of rights, perhaps to rival ideas about which individuals have rights, perhaps to alternative attributions of context-independent rights, perhaps to divergence about which rights that normally accrue can be suspended in the context of inquiry. (Here I have in mind the idea that, as with war or politics, a distinctive "public morality" might apply to scientific investigations.) Imagine, then, ideal deliberators exchange information about the putative bearers of rights and the strategies of inquiry those rights are supposed to debar, and they attempt to defend their conceptions by appeal to principles. In response to such interchanges, they

4. There are several ways of treating this problem formally. For my present purposes, I suppose that the arbitrator proceeds in the way we take disinterested, intelligent, and reasonable people to make their judgments when confronting divergent "expert" opinions.

modify their ideas about which moral constraints are appropriate. At the end of the process, they may find themselves in one of three situations: consensus, no consensus but agreement on one view as the fair representation of different points of view, no consensus and disagreement about how to represent the collective attitude. In the last case, once again, the issue is settled by majority vote.

No doubt there are possible societies—very likely actual societies—in which even ideal deliberation would end in irreconcilable disagreement. People might well fail to reach accord on permissible experimental procedures because they harbor different views about the moral status of animals or fetuses. Even when disagreements persist, the awareness of the sources of divergence in opinion, among parties committed to mutual respect, can affect the ways in which decisions about inquiry are made. So, if it's known that a minority favors more stringent constraints than the majority demands, that fact can be weighed in the choices of a distribution of resources to scientific endeavors, and even in the formulation of the collective wish list. If the deliberators see that a particular outcome could only be achieved by using methods the minority would view as impermissible, they may respond by decreasing the value assigned to the outcome or by committing extra resources so that a strategy acceptable to all may be pursued.

Let's now turn to the third phase of inquiry, the translation of results into applications. Note first that some of the achievements are likely to yield epistemic benefits, to contribute to answering questions that spring from human curiosity; the only issue that arises about these achievements is how to disseminate them both among the scientists to be educated in the next generation and among the wider public. With respect to the achievements that have practical significance, it's natural to think the decision has already been taken in the framing of the research agenda, and the appropriate procedure is simply to follow the policies instituted there. That, however, would be to overlook the possibility that changes in the significance graphs, the consequences of the research that has been undertaken, may modify judgments of relative significance. So I suggest that the ideal procedure at this stage is to mimic the decision-making process of the first phase, with the emphasis now on garnering specified practical benefits. In light of the new knowledge, our ideal deliberators revise their collective wish list, the experts update their views about the probabilities of satisfying various wishes, the arbitrator offers a set of options for gaining particular benefits at various levels of cost, and the ideal deliberators pick a policy for making use of the new information. We may think of that policy as reflecting their newly tutored collective wishes.

The question with which we began—Under what conditions is the science of a society well-ordered?—can now be answered. For *perfectly* well-ordered science we require that there be institutions governing the practice of inquiry within the society that *invariably* lead to investigations that *coincide* in three respects with the judgments of ideal deliberators, representative of the distribution of viewpoints in the society. First, at the stage of agenda-setting, the assignment of resources to projects is exactly the one that would be chosen though the process of ideal deliberation I have described. Second, in the pursuit of the investigations, the strategies adopted are those which are maximally efficient among the set that accords with the moral constraints the ideal deliberators would collectively choose. Third, in the translation of results of inquiry into applications, the policy followed is just the one that would be recommended by ideal deliberators who underwent the process described.

Perfectly well-ordered science is surely too much to hope for. What we would like is, I suggest, a feasible approximation. In setting up structures for the funding and oversight of research, contemporary affluent democracies try, in rather haphazard ways, to come reasonably close to an important ideal. I propose the notion of perfectly well-ordered science as the ideal at which they are aiming.[5]

Before proceeding further, it will be useful to leaven the rather abstract presentation of my answer with some clarifications and illustrations. First, just as I absolved vulgar democracy of the charge that it required actual voting on scientific projects, so too there's no thought that well-ordered science must *actually institute* the complicated discussions I've envisaged. The thought is that, however inquiry proceeds, we want it to match the outcomes those complex procedures would achieve at the points I've indicated. Quite probably, setting up a vast populationwide discussion that mimicked the ideal procedure would be an extraordinarily bad idea, precisely because transactions among nonideal agents are both imperfect and costly. So the challenge is to find institutions that generate roughly the right results, even though we have no ideal deliberators to make the instantaneous decisions we hope to replicate.

Second, like vulgar democracy, the ideal procedure attempts to incorporate the views of every member of the pertinent society. It's an open question as to whether the collection of ideal deliberators contains distinct idealized representatives for each citizen or whether we can assume that people divide into groups whose members are sufficiently similar that they can be represented en bloc. In the latter case, we can suppose that the ideal deliberators proportionally represent the groups with shared perspectives (that is, if one group has twice as many actual members, then it has twice as many ideal representatives). The procedure I've outlined is indifferent as to whether we suppose one-to-one representation or proportional representation of groups with a common perspective.

A third obvious worry about my ideal is its dependence on the values, possibly quite erroneous, of particular societies. This is a direct consequence of my

5. As I suggest in the next chapter, much of the literature on science policy has been handicapped by any clear recognition of what the intended ideal might be.

decision, at the end of the last chapter, to retreat from giving an objectivist answer. Yet, it's natural to think that the only acceptable normative perspective is one that doesn't make science hostage to current beliefs about what things are worth pursuing. My conception of well-ordered science can easily be seen as implicitly recommending that inquiry ought not to lead us to improved views about what is valuable, a prospect many thinkers have taken to be important and liberating.

In response, various things need to be pointed out. First, it's a familiar fact that we can often appraise an activity from either of two perspectives, one that probes its actual success and the other that considers whether an agent did as well as he could, given the limitations of his view. My normative notion of well-ordered science belongs to the latter family. Moreover, as is apparent from my description, the construction of the collective wishes from the individual preferences involves, from the beginning, reflective transformation of those preferences, so there's no danger of holding inquiry hostage to capricious and irrational desires.

Further, if there is indeed a defensible version of objectivism, then it shouldn't be hard to see how to move from my conception of well-ordered science to something stronger. Let's say that the science of a society is well-ordered in the weak sense if it conforms to my criterion, well-ordered in the strong sense if, in addition, its collective values conform to the objective good. When science is well-ordered in the weak sense but not in the strong, then something is amiss. The error is, quite properly, traced to a failure to recognize what's objectively worth doing. But, to recapitulate the point of the last paragraph, the society is still organizing inquiry as well as could be expected, given its limited understanding of the objective good.

Nonetheless, the worry is quite right to spot a danger of conservativism if the only standard for appraising the practice of inquiry is by appeal to my notion of well-ordered science. Conformity to actual values might foreclose investigations that would reveal prejudices and transform aspirations. We'll confront this issue later, in chapters 12 and 13.

A different criticism of my relativization to societies charges that my treatment implicitly focuses on the wrong group. I have written throughout of the practice of science "within a society" and have conceived of the decisions of well-ordered science as representing the wishes of members of that society. The natural interpretation is to suppose that the societies I have in mind are the affluent democracies in which most scientific research is done, and that, for a particular democracy, well-ordered science requires conformity to the idealized wishes of the citizens of that democracy. An obvious defense of that way of proceeding is to invoke the idea that the resources to be committed are those of the society in question, and ultimately of its citizens, so that the citizens have a special right to say how such resources ought to be distributed. Both the interpretation and defense are vulnerable to charges of myopia. Can we really overlook the fact that the kinds of inquiries undertaken have an effect on the well-being of billions of people outside the society? A decision to pursue a line of inquiry that could help in treating diabetes (say) might foreclose opportunities for malarial research. Contemplating examples like this, one can easily conclude that the appropriate group to be represented in the ideal deliberation isn't the citizenry of a particular society (for example some rich democracy) but the entire human species.

We should distinguish a number of positions. One extreme takes the form of the ideal deliberation to be a process that represents only the citizenry of a particular democratic society from which resources will be drawn to pursue inquiry and requires the ideal deliberators to focus only on the needs and aspirations of other members of the society—like the imaginary family, they restrict their attention to one another. At the opposite pole, we can envisage a similar process involving representatives of all members of our species. Two intermediate views are also worth considering. One would continue to restrict membership in the group of ideal deliberators to representatives of the citizens of the society which is to support the inquiry, but would require them to take the preliminary step of acquainting themselves with the needs of people who belong to different societies. In effect, there would be an extra step in the process of ideal deliberation, so that the exchange of views about priorities would include representatives from groups not represented at other stages of the decision. A second possibility is to broaden the class of deliberators to include representatives of other groups whose preferences and opinions count at all stages of the process.

The issues here are complex, and I shall be brief and blunt. Neither of the polar positions seems defensible. Although one might argue that the decisions about inquiry should be left entirely to those who will support it—so that the citizens of an affluent democracy have the right to declare how their funds and talents should be employed—that surely doesn't entail that it's permissible for them to ignore the plight of outsiders, especially when we reflect that their ability to dedicate some of their resources to inquiry may stem from accidents of their society's history or even from past injustices towards those whose priorities are now being excluded. On the other hand, the ideal of a deliberation among parties who cannot be expected to share common democratic ideals or to view one another as participating in joint enterprises looks hopeless: either the tutoring will be entirely inadequate to engender common understanding, or it will effectively transform the ideal deliberators so that their ability to represent the outsiders is highly suspect. Thus I favor one of the intermediate options, and suggest the better choice is that which restricts membership to representatives of the citizens but requires them to become acquainted with the preferences of others, on the grounds that this accommodates a wider spectrum of viewpoints but allows for a shared democratic framework among the deliberators. We might note that, in my original formulation, the ideal deliberators are supposed to take thought for future generations and to consider the implications for members of their society (yet unborn) who will reap the consequences of lines of inquiry now set in motion; by the same token, we might think that the deliberators can and should have an understanding of the consequences for people outside their own society, and it should figure in their deliberations even though the outsiders (like the unborn) do not vote. Clearly, however, much more needs to be said, and I don't pretend to have provided a compelling defense of my preferred way of developing the normative standard.

The last objection I'll consider here is the criticism that the standard for well-ordered science is toothless. Virtually anything, it may be suggested, can be approved. But although well-ordered science might plausibly countenance a number of courses of inquiry, it surely will not allow all. Investigations that bring large benefits to one segment of the society while harming others will not measure up to the standard. Neglect of practical concerns in favor of a sole focus on the epistemic will typically fall short—as will an exclusive concern with the practical (except, perhaps, when practical needs are extremely urgent). The more interesting challenge is to try to understand how far the current practice of the sciences lies from well-ordered science. Can we tell how well (or how badly) we are doing? Can we use the ideal of well-ordered science to improve our situation? To these questions I now turn.

Had we but world enough and time, we could follow a direct approach to designing an ideally well-ordered science. We would review all possible institutions, all possible contexts over which they might operate, formulate an optimization problem, and solve it. This is an impossible dream. We have no realistic prospects of canvassing social institutions and reviewing their entire range of effects across all the situations in which they might be employed—indeed, I think it likely that, in order to assess those effects, we'd already have to resolve many of the issues for which we hope to design inquiry. But we can still scrutinize our own practice from the perspective supplied by the standard.

Consider, for example, the actual ways in which research agendas are constructed. The channeling of research effort is subject to pressures from a largely uninformed public, from a competitive interaction among technological enterprises that may represent only a tiny fraction of the population, and from scientists who are concerned to study problems of very particular kinds or to use the instruments and forms of expertise that are at hand. Actual deliberations (as we'll see shortly) often involve agents who depart from the ideal in two different ways: potential consumers who have a highly incomplete understanding of the range of options and of their consequences, and inquirers who are strongly motivated to present research projects in ways they think will appeal to a much broader public.[6] So, in this context at least, there are grounds for pessimism.

Contrast the appraisal of agenda-setting with that of the moral constraining of inquiry. Here, from the relatively mundane and unperturbing potential transgressions (scientific fraud, plagiarism, and so forth) to the truly disturbing cases (experiments that damage human subjects without their consent), we can point to a core set of moral constraints that are close to being universally acknowledged, and serious attempts, at least, are made to ensure that researchers abide by them. Partly because of terrible abuses of inquiry in the past—the examples of the Nazi doctors and the Tuskegee experiment—widely shared views about human rights have inspired systematic oversight of experimentation involving human subjects, which quite deliberately involves people with a variety of perspectives. Further, when members of contemporary societies hold fiercely opposing attitudes, as with research on human embryonic tissues or the use of nonhuman animals, the existence of a lively debate about the moral standing of the pertinent entities has created fora that are plainly intended to approximate something like the types of deliberation and negotiation I've described.

In what follows, I'll attempt to identify some likely problems for our current practices, thus indicating loci where we might hope to do better. I'll start with three different types of concern about the setting of research agendas and the use that is made of scientific results: one charges that the preferences of large segments of the public are consistently neglected, a second alleges that inquiry is distorted because the untutored preferences of outsiders lead to the neglect of problems of real epistemic significance, while a third suggests that the coherent systematization of widely shared preferences would recommend different priorities. The first often takes the form of complaints that the sciences don't take into account the needs of women, children, members of minorities, and people in developing countries; the second typically comes from scientists who have been disappointed by the lack of support for a project that fascinates them; the third usually comes from those who oppose a scientific project on the grounds that a systematic interest in the values professed by the champions of the project would lead to quite a different assignment of resources. There's no doubt the actual processes that shape our research agenda and convey the results of inquiry to the public give disproportionate emphasis to the predilections of people belonging to particular subgroups while members of other subgroups don't

6. Although there are obvious instances in which this discrepancy arises, it would be wrong to maintain that the interests of scientists are always antagonistic to those of the broader public. Part of Donald Stokes's thesis in *Pasteur's Quadrant* is that considerations of *perceived* epistemic significance and *perceived* practical import may sometimes coincide. Even here, however, there may be a deeper problem. For the scientists are responding to their perceptions of what society needs, and that may not coincide with the outcome chosen after the ideal deliberation I've outlined.

participate directly at all. Yet one might think that the agenda and the applications must be sensitive to a wide variety of preferences because at least some people who do play a direct role in the decisions—administrators of government agencies, manufacturers, and other entrepreneurs—are answerable to the public. Optimists hope there will be some type of invisible hand, so that apparently unrepresented minorities, however small, can offer a niche to which businesses and politicians will want to appeal, and thus affect the character of inquiry. If a practical problem is urgent for a subgroup then, within the public sphere of assigning resources to inquiry, elected representatives should find it advantageous to encourage research that addresses the problem, and, within the private domain, there will be room for commercial exploitation. Unfortunately, it's very easy to show that there are conditions under which a set of rational agents—whether bureaucrats or entrepreneurs—will do better to ignore the problems of small minorities: if the distribution of constituencies across the voting population assigns each electoral district a dominant group interest while the minority is thinly spread, and if the costs of starting to develop the pertinent technology are sufficiently high, the minority won't be worth bothering about. Further, given the account of the evolution of inquiry that I have offered, there are further reasons for thinking that an initial decision to favor the interests of one group may be self-perpetuating. A line of solution to a practical problem, and its associated research projects, may be suboptimal for a subgroup of the population *relative to a class of options that were never offered*, even though it's the best of those that continue to be available, precisely because of an original neglect of the preferences of the subgroup.

The root idea is that a decision to extend the significance graph in a particular direction may make it easier to continue in the same direction, perhaps by decreasing costs, perhaps by increasing chances of success. I'll illustrate it with a popular (but controversial) example, the dedication of resources to devising effective means of birth control. Let's assume that, prior to the research that led to the Pill, men would have preferred a pill that could be taken by women, women a pill that could be taken by men.[7] Suppose further that the initial decision ignored the preferences of the large majority of women. For many people, men and women, avoiding conception in sexual intercourse is an important goal, and hence the provision by biotechnology (*avant la lettre*) of the female pill seemed an excellent solution—at least until there were concerns (possibly unjustified) about increasing rates of cancer and heart disease. Yet, given our assumptions, women were offered a choice that didn't completely reflect their

7. This strikes me as a lot more complicated than is often supposed, primarily because of issues about control. Of course, preferences have evolved in light of changing sexual mores and because of the frequency and severity of sexually transmitted disease.

preferences, the female pill vs. much cruder forms of contraception, when they might have been given the choice of the female pill vs. the male pill. Optimists think that, once the broader spectrum of possibilities is recognized, there will be pressure on inquiry to respond. But the invisible hand fails. Because the female pill has a head start, even if *both* projects are now pursued with roughly equal resources, it will be expected that the choice will be between a female pill with costs C (measured in terms of side effects as well as money) and a male pill with costs C^+ (where, as the notation suggests, $C^+ > C$). Even if women prefer a male pill to a female pill, with equal costs, they may continue to prefer a female pill at lower cost to a male pill at higher cost. There's a *Nonrepresentational Ratchet*: because of the initial neglect of female preferences, women never receive the choice they want.

The story I've sketched may be right—or it may not. Detailed sociological work is needed to decide that issue. My aim at this stage is to canvass possibilities, and the first of these is

> *The Problem of Inadequate Representation*
>
> A group is inadequately represented when the research agenda and/or the application of research results systematically neglects the interests of the members of that group in favor of other members of society. Because of the Nonrepresentational Ratchet an early problem of inadequate representation in a field may be self-perpetuating.

One can't show that the problem of inadequate representation exists simply by noting that members of a particular group aren't sufficiently represented in decisions about inquiry—but, by the same token, nor can one suppose that some invisible hand will operate to forestall the problem.

The second problem I'll consider stems from the fear that representation of perspectives outside science works too well. Because the preferences of the vast majority of citizens are untutored, areas of science that depend heavily on public funding can be shaped by governmental decisions that respond to widespread ignorance, with the result that practical projects whose significance can easily be appreciated are overemphasized, with concomitant neglect of questions of large epistemic significance. Although there are public discussions about budgets for scientific research, discussions that sometimes afford inquirers an opportunity to campaign for their favorite epistemic endeavors, these may be quite inadequate to solve the problem. Despite the compelling testimony scientists present to elected representatives, those whom they hope to convince are responsible to constituents whose preferences run strongly counter to the appeal for funding. The mirror image of the problem of inadequate representation is

> *The Problem of the Tyranny of the Ignorant*
>
> Epistemically significant questions in some sciences may systematically be undervalued because the majority of members of society have no appreciation for the factors that make those questions significant.

Once again, it's important not to conclude too quickly that we are confronting an instance of this problem. Just because scientists don't succeed in obtaining the resources required for their favorite projects, it doesn't follow that we aren't in a state of well-ordered science—after all, even if the public preferences were tutored, they might still oppose the line of inquiry envisaged.

Scientists' awareness of potential problems of the tyranny of the ignorance spawns a further chance of departing from the state of well-ordered science. Favored lines of inquiry can be promoted by advertising them as catering to the wishes of large segments of the community. Even if it should turn out that citizens' tutored preferences would fortuitously accord with the recommended agenda and the envisaged applications, misleading accounts of what can be expected introduce further political constraints on research (as scientists must make gestures at accommodating the expectations they have raised) and can also reinforce attitudes that oppose research responsive to tutored preferences. In a nutshell, public misperceptions of the rationale for research and applications don't foster the stable pursuit of inquiry that would correspond to tutored preferences.

The defenses of the genomes project exhibit very clearly why this is so. In some quarters, the important reason for mapping and sequencing the human genome is taken to be the opportunity to protect American leadership in biotechnology. Public testimony to Congress on behalf of the project, widely reported in the media, presented a very different line, emphasizing the biomedical breakthroughs that were likely to occur. Partly because the project's champions sometimes said as much, partly because it was what the public expected, the biomedical advances were understood in terms of readily achievable strategies for the prevention and treatment of disease. But, as I noted in chapter 1, a sober review of the relief afforded by enhanced understanding of the molecular bases of diseases offers a very mixed picture, with a few partial successes and some cautionary failures. It was not pointed out to Congress, or to the general public, that the immediate practical consequence of mapping and sequencing would be an enormously enhanced ability to offer genetic tests, typically without being able to give much advice for addressing health risks: that this would be likely to reveal painful information to patients, especially in a probable majority of situations in which genetic counselling would be ineffective; that it might well serve as the basis for new forms of discrimination; and that it would result in the proliferation of prenatal tests, which in turn could be expected to multiply, possibly by a significant factor, the number of abortions.

For most scientists there has always been a far deeper motivation. Developing sequence technology and applying that technology to selected nonhuman organisms is expected to make biologists of our century better able to explore the large questions of physiology, developmental biology, and even evolution (genomic analysis will shed light on evolutionary relationships and reveal the kinds of changes involved in speciation). The scientists involved believe that the work in which they're engaged will ultimately translate into an enormously richer and more complete view of physiology and development—although they would concede that this is likely to take a very long time—and that there will be consequent medical benefits.

This is eminently justifiable and probably correct. The envisaged strategy is akin to that adopted by early twentieth-century geneticists who self-consciously sought to resolve the most fundamental issues by working with tractable organisms rather than tackling the questions of human medical genetics head-on. If the significance graphs for the pertinent fields were clearly articulated and their historical development explained, it is quite possible that the wisdom of the strategy would be evident and that the inquiries envisaged would accord with the tutored preferences of the citizens whose taxes support the genomes project. Although the research agenda actually pursued adequately represents the preferences people would acquire as the outcome of ideal deliberation, the way of achieving this goal is unreliable, and that unreliability has serious consequences. Funding flows because biotechnology is viewed as a continuing source of jobs for Americans and because medical benefits are believed to be around the corner. The entrenchment of these beliefs causes trouble for inquiry because there's pressure to produce *some* kind of short-term "solutions," and it also distorts the applications of results by concealing the social problems that the real products of the project are likely to bring. After all, who needs to worry about inadequate counselling, lack of insurance coverage, and genetic discrimination when cures are just around the corner? Hence we currently confront

> *The Problem of False Consciousness*
>
> A research agenda may conform to the tutored preferences of the majority not because the public reasons for the agenda are those that would figure in an ideal deliberation, but because those reasons misrepresent the agenda in ways that cater to the actual (untutored) preferences of the majority. Because these preferences are not tutored, there may be harmful constraints on the pursuit of inquiry and serious threats to the proper application of its results.

Faced with the prospect of the tyranny of the ignorant, false consciousness may provide a way of reaching a better outcome. Yet it's not a feature of well-ordered science.

The last general problem I'll discuss has to do specifically with the applications of research. Sometimes a course of inquiry can be defended and publicly supported because its champions advertise its promotion of a goal that is highly valued in the community, even though it's not made clear (quite possibly because the advocates don't recognize the point) that a different course of inquiry or application would promote that value more completely or more justly. This problem, too, can be exemplified by the genomes project. One important reason for mapping and sequencing the human genome is the possibility of preventing the births of people who would suffer from devastating genetic diseases—the extension of the benign programs of prenatal testing begun in the attack on Tay-Sachs disease. Ideal deliberation of the scientific research agenda and of the translation of scientific knowledge into technology and public policy would thus invoke the goal of diminishing the number of children whose lives are doomed to have abysmally low quality.

But once this goal is recognized, as it should be, then our ideal deliberators ought to consider all the available ways of advancing toward it, reviewing possible projects that might improve the quality of children's lives, especially in those instances where we know that the expected quality, without intervention, is low. Consider now the fact that something of the order of a million American children live in apartments where they are exposed to toxic levels of lead. There's no current program for clearing up their environments and preventing them (and future children) from suffering severe consequences (sometimes damage to bodily organs, sometimes impairment of mental function). An obvious criticism of our actual practice of extending and applying science is that our research agenda would be ideally supported by a principle that would also favor the application of technologies we already have (techniques for removing lead) and inquiries designed to develop new technologies to fulfill the same function more efficiently. In other words, we face

> *The Problem of Parochial Application*
>
> An actual research agenda and a practice of application may be ideally supported by a principle that would licence forms of research not currently undertaken or applications of previous research that are not pursued.

Parochial application often occurs because it goes hand in hand with inadequate representation. In the genetic example, it seems plausible that the failure to apply the principle of improving the well-being of children by launching a program of lead removal results from neglecting the interests of people who are likely to live in the pertinent inner-city dwellings, to wit, members of ethnic minorities.

I've delineated some general problems in an effort to show that, even without detailed optimality analyses, we can sometimes identify ways in which the practice of the sciences is likely to diverge from well-ordered science.[8] I'll close with a more detailed discussion of whether we would be better off if there were more public input into decisions about which inquiries should be pursued.

An obvious thought is that involving representatives of diverse perspectives in decisions about prospective inquiries and about the uses of existing knowledge would be likely to modify the aims pursued to bring them closer to those that would emerge under conditions of ideal deliberation. Contrast a number of ways in which research agendas might be set. One, *internal elitism*, consists in decision-making by members of scientific subcommunities. A second, *external elitism*, involves both scientists and a privileged group of outsiders, those with funds to support the investigations and their ultimate applications (call these people "paymasters"). A third, *vulgar democracy*, imagines that the decisions are made by a group that represents (some of) the diverse interests in the society with advice from scientific experts. The fourth, *enlightened democracy*, supposes decisions are made by a group that receives tutoring from scientific experts and accepts input from all perspectives that are relatively widespread in the society: in effect, it fosters a condensed version of the process of ideal deliberation I've outlined.

I take it that the status quo in many affluent democracies is a situation of external elitism that groups of scientists constantly struggle to transform into a state of internal elitism. Vulgar democracy is, as I've insisted, likely to be a bad idea. The interesting question is whether enlightened democracy would be preferable to either form of elitism.

There are three influential arguments that incline people to dismiss the possibility. The first, already hinted at above, invokes the idea of an invisible hand. Consider the incentives for paymasters under external elitism. To achieve their ends they must respond to the preferences of the main constituencies within society, and their decisions must thus take into account the heterogeneous preferences of the citizenry. Explicit representation of those preferences isn't needed, and may well prove inefficient.

Earlier, I suggested we have no reason to believe in the invisible hand. Here, I want to add three points. First, even if the paymasters' decisions respond to the preferences of citizens, they won't reflect the transformation of those preferences that would occur under tutoring. It's thus unlikely that the agenda will be

8. In my judgment, I've only noted the most obvious problems. Once the ideal of well-ordered science is recognized, there's an important need for a political theory of science that will consider the various ways in which the interests of actors and social institutions might easily divert us from the outcomes that would be reached in a state of well-ordered science. In *Between Politics and Science* (Cambridge: Cambridge University Press, 2000), David Guston offers a framework that might be valuable in developing a theory of this kind, although, as with many other writings in this area, Guston's work seems to me to suffer from a lack of the ideal of well-ordered science.

set in a way that even approximates the best pursuit of inquiry. Second, even if there were some pressure to respond to the untutored wishes of major constituencies within the society, we can expect minority concerns will be slighted —even in those instances in which ideal deliberation would have produced a response to them. The failure to represent the interests of people beyond the society is likely to be even more severe. Finally, it's highly likely that paymasters will prefer to manipulate the preferences of those to whom the products of inquiry are to be offered in whatever directions will maximize profits, and there's every reason to believe that these will not coincide with the products of ideal deliberation. Reflections on the current biotechnology market in the United States are hardly encouraging.

The second argument in favor of elitism suggests that introducing citizens into the deliberations that shape research agendas would reduce the pool of available strategies by making it less attractive for paymasters to support research and for talented people to engage in scientific careers. Here we encounter familiar advertisements for the benefits of deregulation. Yet the fact that representatives of citizens with different perspectives would be involved in discussions about agenda-setting and about applications hardly affects the prospects of paymasters; indeed, under many circumstances, those who underwrite research and development are happy to pay for information about the constituencies to whom they intend to appeal. Perhaps, however, there are disincentives for the scientists whose views about what lines of inquiry are most interesting might be swamped by the demands of the outside majority.

Here too the existence and strength of the disincentive could be questioned, for it's worth recalling that brilliant people are sometimes prepared to spend their lives in carrying out the research projects directed by commercial (or governmental) concerns. More important, the argument assumes, without justification, that there's no way to organize broader deliberation so as to tutor the preferences of all discussants, generating a research agenda acceptable both to the discussants and to those who are to carry out the research. What lurks behind the suspicion is, I think, the thought that any attempt at democracy must be sufficiently close to vulgar democracy that the tyranny of the ignorant will be inevitable.

This is the last, and I think the most powerful, reason for defending elitism. Enlightened democracy would try to tutor the raw preferences of representatives of different perspectives within the society, would admit expression of the needs and perceived interests of all groups, and would thus conduct informed deliberations. Skeptics pose a dilemma: either the processes that precede agenda-setting are impossibly cumbersome and time-consuming or they fail to shift the views of the participants sufficiently to produce a genuine departure from vulgar democracy.

The best response to that dilemma would be to delineate, clearly and specifically, a mechanism for enlightened democracy and to show, on the basis of sociological research and mathematical analysis, that the expected results are better than those that elitism would yield. It's not hard to do the mathematics (indeed, it can be developed in similar ways to those used in formal treatments of the problems of the last chapter), but the sociological information required to build realistic models is currently not available. Hence, I must settle for a weaker response. Democratic proposals within other areas of politics and political philosophy are always vulnerable to charges that the incorporation of preferences would be too cumbersome. We rightly reject *a priori* skepticism until we've explored whether the democratic suggestions can be made to work, either by acquiring empirical information or by trying various possible schemes (possibly on a limited scale). Unless, or until, sociological research shows that the project of approximating tutored collective preferences is hopeless, we have no basis for concluding that some form of elitism must be superior.

Doubts may linger. Perhaps they can be allayed by a brief look at three policy formulations that have had some influence on the institutional arrangements within which the sciences are practiced.

ELEVEN

Elitism, Democracy, and Science Policy

THE FIRST REPORT ON SCIENCE POLICY was written as a fable. In *New Atlantis*, left incomplete at his death, Bacon offered a tale about the crew of a sailing ship, who, after various disasters, find a haven in the island of Bensalem. Here the mariners are treated with great hospitality, and they are surprised by the wisdom, generosity, and incorruptibility of the island's government. Liberality and fair-dealing are founded on the institution of an elite group of investigators, the members of Salomon's House, who seek "the knowledge of causes, and secret motions of things."[1] In the terms I've been using, they aim to achieve both epistemically and practically significant truths, and they are very clear about the character of both.

Bacon took for granted a substantive conception of the kinds of things about which people are curious, essentially a late Elizabethan/Jacobean vision of the "wondrous phenomena" of the natural world. Equally, he offered a substantive conception of human needs, so that the caves, chambers, storehouses, dispensatories, and workshops of Salomon's House respond to the requirements of nourishment, health, warmth, protection from the elements, and other things he saw as being universally desired. There's no suggestion that the priorities and preferences of Bensalem's citizens might differ, nor that choices might have to be made among the fruits of inquiry. Further, the story betrays no hint that the "useful inventions" might become the property of particular individuals or groups within the society, so that people could advance their own interests by trading in them.

The work of Salomon's house is carried forward by coordinating au-

1. Francis Bacon, *New Atlantis* (Oxford: Oxford University Press [World's Classics], 1966), 288.

tonomous decisions made by the members in discussion with one another. At various stages, they confer to plan the next steps to be taken (following the method for individual inquiries Bacon outlined in many places). The fellows decide what to do, what to publish, what to keep secret, and what applications to make. So we are offered an explicitly elitist vision of well-ordered science, one that takes an objectivist vision of the good at which inquiry aims. There are certain things which it is good for human beings to know because it will relieve their curiosity, certain things which it is good to know because applying the knowledge will contribute to human welfare. The wise inquirers understand these things and thus bestow all kinds of benefits on their society.

It's evident from the early documents about the Royal Society that Bacon was a tremendous inspiration to whose who created institutions for the practice of the sciences in Britain. Residues of his elitist conception of well-ordered science linger, not only in the English-speaking world, but in other democratic societies. Yet it's highly doubtful that we'd accept either Bacon's own vision of what is good for people or even the idea that the wise experts can be expected to know what's objectively in human interests. Pluralistic democracies are thoroughly attuned to the possibilities of individual differences that affect well-being, and, after Mill, it's hard to countenance the view that people should have no role in deciding what is worthy of being pursued. Unlike Bacon, we recognize that inquiry might generate a mixture of beneficial and harmful consequences, and that these may be distributed in ways that are biased or unfair.

So it's hardly surprising that later policy proposals are not as unself-consciously elitist as Bacon's. But, as I shall suggest, even the most thoughtful of them don't articulate the democratic ideals as clearly as we should require.

The most important document about the place of scientific research in a twentieth-century democracy is surely Vannevar Bush's *Science—The Endless Frontier*. At the close of the Second World War, Bush responded to a request from President Roosevelt to advise on how scientific research might be used for "the improvement of the national health, the creation of new enterprises bringing new jobs, and the betterment of the national standard of living."[2] Bush's letter of transmittal contained a significant addition to the President's wording. It closed with the sentence: "Scientific progress is one essential key to our security as a nation, to our better health, to more jobs, to a higher standard of living, and to our cultural progress."[3]

Bush's report was masterly in combining two perspectives that are hard to reconcile. On the one hand, he and the scientists who wrote on specific aspects

2. This passage occurs in Roosevelt's letter to Bush of November 17, 1944. See Vannevar Bush, *Science—The Endless Frontier* (Washington D.C.: NSF, 40th anniversary ed., 1990), 3.

3. Bush, *Science—The Endless Frontier*, 2.

of the organization of research wanted to insist on the value of inquiry for a wide range of public concerns. On the other hand, they wanted to protect science from outside direction and interference, arguing that "basic research" should be free, that it should be given "special protection and specially assured support."[4] Their most popular rationale emphasized that "basic research" is crucial to avoid stagnation. In a telling metaphor, Bush saw such research as providing "scientific capital" on which future technological ventures can draw.[5] Less prominently, the report contained hints of the intrinsic value of scientific discoveries—recognition of epistemic as well as practical significance—in Bush's reference to "cultural progress" in the letter of transmittal and sometimes more explicitly: "Moreover, it is part of our democratic creed to affirm the intrinsic cultural and aesthetic worth of man's attempt to advance the frontiers of knowledge and understanding."[6]

The arguments of Bush's report were designed to combat an alternative view, vigorously defended by a West Virginia senator, Harley Kilgore.[7] On Kilgore's vision, federal funding for research would be more directive, addressing social problems and attempting to distribute the benefits of science and technology equitably across the population. The challenge for Bush and his colleagues was to square their claims that scientific research would provide many advantages for the entire nation with the elitist view that scientists were the best judges of how such advantages should be provided, and, in particular, that their disinterested pursuit of "fundamental questions"—presumably issues that strike them as having greatest epistemic significance—would be the best policy for long-term results.

In effect, they did not move far from Salomon's House. Bacon's list of goods to be pursued gave way to an updated version: we want national security, relief from disease, more jobs, technological gadgets that save time and labor. Perhaps Bush might have been prepared to admit that this list of desirable consequences should be refined, even substantially modified, through extensive discussion in which outsiders would express their points of view. For that would hardly have mattered to the main thrust of the report. The constant refrain is that *whatever* people want from research, attention to "basic research" is the most effective way of providing it: "Scientific progress on a broad front results from the free play of free intellects, working on subjects of their own choice, in the manner

4. Bush, *Science—The Endless Frontier,* 83.

5. Bush, *Science—The Endless Frontier,* 19; see also 18, 21–22, 81–83.

6. Bush, *Science—The Endless Frontier,* 79.

7. For an informative discussion of the clash between the programs of Bush and Kilgore, see Daniel Kevles, "Principles and Politics in Federal R&D Policy, 1945–1990: An Appreciation of the Bush Report," printed as an introduction to the 1990 edition of *Science—The Endless Frontier.* See also Kevles, *The Physicists* (Cambridge, Mass.: Harvard University Press, 1971), chap. 22, "Victory for Élitism."

dictated by their curiosity for exploration of the unknown."[8] Bush and his colleagues thus defended an extremely strong claim. However issues about the genuine interests of American citizens are settled, and whatever the best way of applying scientific research to promote those interests, supporting scientific research along whatever lines the scientific community thinks are most promising will provide the best way for inquiry to address those interests.

Bush had no detailed empirical studies of inquiry under different conditions of organization from which he could draw to support his claims. Indeed, the necessary studies are still lacking. Moreover, he had to cope with a salient counterargument. After all, the war effort had effectively organized the nation's scientific talent and directed research towards immediate practical needs. Why, then, should one think that elitism is necessary for fruitful inquiry? Bush's report answered the question in two ways. First, it proposed that the success of directed research in the war years depended on "a large backlog of scientific data accumulated through basic research in many scientific fields in the years before the war."[9] To continue to pursue directed research in times of peace would thus be the scientific equivalent of eating the seed-corn. Second, in an explicit reply to the "very natural" thought that "peacetime research would benefit equally from the application of [planned coordination and direction]," Bush's advisory committee on Science and the Public Welfare (the Bowman committee) wrote:

> Much of the success of science during the war is an unhealthy success, won by forcing applications of science to the disruption or complete displacement of that basic activity of pure science which is essential to continuing applications. Finally, and perhaps most important of all, scientists willingly suffer during war a degree of direction and control which they would find intolerable and stultifying in times of peace.[10]

It isn't hard to find the intended conclusion in these passages, but it's extremely difficult to discern any argument for that conclusion. The report simply asserted that "pure" research is essential for continuing technological health (how fortunate that the war didn't drag on for a few more years!) and gestured at the feelings of scientists, as if people were incapable of concentrating their efforts on a variety of cooperative projects and as if whatever discontents flowed from the direction of research would be inimical to the flourishing of inquiry.

As the report moved away from the problematic example of the war effort, there were clear attempts to do better than bare assertion. The Bowman committee remarked that scientific discoveries often "come as a result of experi-

8. Bush, *Science—The Endless Frontier*, 12.
9. Bush, *Science—The Endless Frontier*, 13.
10. All citations from Bush, *Science—The Endless Frontier*, 80.

ments undertaken with quite different purposes in mind."[11] They suggested that, because of urgent pressures to solve practical problems, "*applied research invariably drives out pure,*"[12] and offered a brief history of public funding of American science in the nineteenth century which tried to link economic progress to the investment in pure research. None of these discussions, however, yields a cogent argument for elitism. Even if the authors had offered a less impressionistic history, it would be extremely difficult to connect the transformation of the American economy in the nineteenth century to the pursuit of "pure research." The emphasis on the serendipity of discovery, which sees scientific breakthroughs coming in unanticipated ways, hardly demonstrates that what a group of scientists views as the hot topics for "basic science" will yield the means for satisfying the desires of a wider public. Strictly speaking, all that such examples show is that inquiry can lead to unexpected destinations, but the conclusion the authors wanted to draw is that one way of choosing a direction is particularly likely to be auspicious. Finally, the worry that pure research will always prove vulnerable in competition with practical projects seems to derive from the thought that the alternative to elitism is a benighted form of vulgar democracy. The report did not envisage the possibility that decisions about the future course of research initiatives might be made by reflective people who had absorbed whatever the historical record can reveal about the latent value of "basic science" and recognized the dangers of focusing on practical payoffs alone.

In the end, it's hard to see how the Bush report could have defended its elitist claims in the absence both of serious empirical evidence and of a fully articulated ideal. Although the authors attempted to discuss how the sciences relate to public welfare, they had no detailed conception of the well-being that inquiry should promote. In particular, there was no thought that generating some kinds of epistemically significant truths might be in the interests of a broader public, and that the public, reflectively aware of the possible extensions of inquiry, might avoid the myopia which the report viewed as its characteristic condition. As a consequence of these omissions, the institutional framework the report constructed is one in which scientists are left free to pursue their own curiosity but simultaneously saddled with the task of advertising their research as potentially satisfying the untutored preferences of the public. It's easy to see how the stage was set for later struggles (only partly acknowledged) between scientific efforts to manipulate public funding and suspicious outsiders.

Despite its elitism, Bush's report, unlike Bacon's fable, was plainly permeated by democratic values. Nowhere is this more evident than in the committee report on scientific education. That committee explicitly rejected the idea that a

11. Bush, *Science—The Endless Frontier,* 81.
12. Bush, *Science—The Endless Frontier,* 83.

scheme for science education could be judged simply according to its capacity for producing the best researchers in the most efficient fashion. The authors realized there were broader political goals to be served: "We think we probably would not, even if we were all-wise and all-knowing, write you a plan whereby you would be assured of scientific leadership at one stroke. We think as we think because we are not interested in setting up an elect. We think it much the best plan, in this constitutional Republic, that opportunity be held out to all kinds and conditions of men whereby they can better themselves."[13] Lurking in the background, there thus seems to be a broader democratic ideal, one that sees the framework for pursuing inquiry as contributing to the welfare of citizens in different ways (for example, by opening careers to "all kinds and conditions of men"), so that the facilitation of "pure science" isn't the be-all and end-all of a science policy. Perhaps at least some of the authors were groping towards the ideal of well-ordered science, but, without articulating it, they couldn't frame the possible options or raise the genuine empirical questions about which is to be preferred. The result was a document in odd tension with itself, a thoughtful and rightly influential brief for science that, for all its virtues, rested on muddled and inconsistent foundations.

By the mid-1990s it had become evident that, at least for some types of research funded by the government, the kinds of procedures favored by the Bush report might be too elitist. In 1998, the Institute of Medicine published the report of a committee, chaired by Leon Rosenberg of Princeton University. Entitled *Scientific Opportunities and Public Needs*, the report aimed to improve priority-setting and public input at the National Institutes of Health. Rosenberg's preface indicated the sources of the problems—a belief that NIH is more concerned with curiosity than care, the frustration of groups that are not heard, a lack of understanding about NIH priority-setting—and he concluded by thanking "the public, who reminded us of the purpose of NIH and of the democratic ideals that must permeate effective stewardship of a federal agency."[14] The discussion that follows is sensitive to many public concerns about how lines of inquiry in the health sciences are selected, but, like Bush's discussion, it is handicapped by its failure to analyze the goal(s) of research in this area.

The Rosenberg committee reviewed previously published criteria for setting priorities: public health needs, scientific quality of the research, potential for scientific progress, portfolio diversification, and adequate support of infrastructure. After an analysis of some of these criteria, the committee indicated its "general support" for them and suggested that their application would be im-

13. Bush, *Science—The Endless Frontier*, 149.

14. *Scientific Opportunities and Public Needs* (Washington, D.C.: NIH, 1998), ix.

proved by obtaining better statistics and by making systematic use of such statistics. Along the way some fundamental points were bypassed.

The first issue, which the committee acknowledged but did not resolve, concerns the relative weight given to the various criteria. Plainly, decisions to give highest weight to "scientific quality" might differ quite extensively from decisions to concentrate on the most serious public health needs. In its recommendations, the committee proposed that the criteria be employed in a "balanced way," but did not specify what this means or what kind of procedure is likely to generate "balanced employment." In the end, then, critics who worry that NIH is far too concerned with satisfying scientific curiosity were not given an account that might allay their qualms.

A second, similar, question focuses on the notion of public health needs. The committee rightly insisted on a "broader view of health as leading a full and high-quality life even in the presence of pathologies, chronic symptoms, and functional limitations."[15] In light of this conception, the authors criticized an earlier NIH proposal that public health needs are judged by the incidence, severity, and cost of specific disorders. Once again, we're plunged into conflicting criteria for judging the relative benefits of possible outcomes with different incidences of different diseases, and, once again, the committee failed to provide a resolution. The report noted, quite correctly, that, without an answer to this question, NIH can only guess the extent to which its ways of setting priorities are effective. Bacon and Bush might have maintained that it's possible to give an objective answer, that one state of disease-incidence promotes human welfare more than another. The ideal of well-ordered science supposes that such states are ranked by considering the tutored collective preferences of ideal representatives, and so might regard the education and canvassing of public opinion as methods of moving policy decisions in the preferred direction. Because the report supplies no specification of the aim, it was, as we'll see, ultimately unable to make a serious case for involving the public in the setting of research priorities.

Before turning to the main theme of the report, namely the need for greater public input, it's worth looking briefly at a seductive argument, offered in support of the idea that the status quo is reasonably good. In appraising past practice at the NIH, the report noted that "the balanced application of these criteria has led to the accumulation of basic knowledge about human biology that is unparalleled in the history of science."[16] Waiving suspicions about what "basic knowledge" means, or how one might measure amounts of it, there's a very obvious alternative interpretation of the history. We're impressed (probably rightly) with the shift between the state of knowledge at the time just prior to the injection of funds into the NIH and the state of knowledge at the present.

15. *Scientific Opportunities and Public Needs,* 31.

16. *Scientific Opportunities and Public Needs,* 38.

Yet, since the funding support was "unparalleled in the history of science," it's genuinely hard to say whether the "balanced application of the criteria" has anything to do with the success. The point is elementary. One can't claim that the use of the criteria has contributed unless there's a "control"—a point on which committee members would surely insist if they were evaluating a scientific proposal for "high quality"—and the history supplies no "control."

Despite their confidence that things have been going well, the committee recommended that the priority-setting processes should be open to more public input. (In light of their failure to articulate an ideal at which inquiry aims, it's not obvious why this would be an improvement: perhaps it's just a response to the fact that some of the lay citizens have been getting restless?) They recognized that the present structure of NIH is the result of past contingencies that have lumped together some disease conditions (or bodily systems) and separated others, and they point out that, as things stand, each institute director receives advice from a committee, one third of whose members are nonscientists (but typically physicians). Improvements could be made, they suggested, by giving the director of NIH a more systematic role in setting priorities and increasing the diversity of the membership of the advisory committees. As the committee saw very clearly, there's a "perception" that "NIH does not encourage public input at the highest levels."[17]

What, then, is to be done? The Rosenberg committee explicitly refrained from trying to replace the "existing criteria for priority setting." Instead, they aimed to suggest mechanisms "through which public voices can be heard in a constructive and open manner," and, to this end, recommended forming a Council of Public Representatives "to facilitate interactions between NIH and the general public" and urged that the public membership of advisory groups "should be selected to represent a broad range of public constituencies."[18] Who could disagree? For all their admirable intentions—and it's notable that the discussion of mechanisms begins by asserting that "Public input is an essential and integral part of any democratic process"—it remains entirely unclear what the envisaged mechanisms of public input will actually do and what they would be expected to do. The committee was quite perceptive in appreciating the need for "two-way communication between NIH and the public," in other words for the tutoring of the preferences of outsiders, but did virtually nothing to address this problem. Similarly, there are hints that the committee saw how public participation in health policy is easily biased by the strenuous efforts of interest groups, some of which succeed in attracting capital and publicity while others do not; again, nothing was done to channel the public input so that the recommended council is not simply an arena in which spokesmen for various narrow perspectives compete for attention.

17. *Scientific Opportunities and Public Needs*, 52; see also 48, 50–51.

18. Citations from *Scientific Opportunities and Public Needs*, 61, 62.

Decision-making is to remain in the hands of the scientists, although the NIH director will receive "valuable and thoughtful perspectives" on research programs from "those who are in some way affected by disease and disability."[19] But *all* of us are affected by disease and disability—most people have a particular interest in preventing disease—and the deliberation about policy priorities ought to range beyond any person's (or group's) narrow short-term interests in response to a clear perception of the scientific opportunities. Perhaps the Rosenberg committee assigned the council only an advisory role—leaving the final choices to the NIH director and other experts—because it feared the dangers of vulgar democracy. If so, then the high rhetoric about the need for public input in democratic processes is in tension with a deep pessimism about what the public can offer. The recommendations have the air of palliatives, designed to appease the critics of NIH, precisely because the committee has failed to articulate just what the democratic ideal should be and because it hasn't addressed the empirical questions about how to promote that ideal.

I have been critical of thoughtful attempts to organize scientific research in a democratic society, undertaken by eminent scholars who have been moved by excellent intentions. My diagnosis of the shortcomings of their efforts in science policy sees them as not having a sufficiently clear goal: they haven't posed the fundamental question, "What is the collective good that we want inquiry to promote?" Only in light of an answer to that question can we see what empirical issues need to be addressed, and, on the basis of resolution of those issues, what policy recommendations might be appropriate. My discussion of well-ordered science is a proposal for answering the fundamental question. It frames empirical issues about how we might arrive at a council in whose discussions tutored preferences were exchanged to form something like the collective wish list, and it is not hard to envisage how social research could make headway with those issues. Instead of a science policy that has swathed a commitment to elitism in attractive drapery, we might succeed in fashioning something genuinely democratic.

During the 1960s, there was a brief period during which scholars concerned with science policy took up what I view as the fundamental question: What exactly is the goal of scientific inquiry in a democratic society? (Or, in economic terms, what precisely are we hoping that science will maximize?) Edward Shils offered a lucid presentation of the problem: it arises from the fact that scientific research is expensive, so that there are too few resources for pursuing all the projects we conceive; accordingly "[h]eads of expenditure have to be ranked in accordance with the intrinsic and instrumental value of the activities to be supported by the expenditure."[20] Shils' distinguished contributors wrestled with the

19. *Scientific Opportunities and Public Needs*, 65.

20. Introduction, E. Shils, ed., *Criteria for Scientific Development: Public Policy and National Goals*, vii.

issue. Michael Polyani claimed that the "Republic of Science" is a "Society of Explorers," striving for "intellectual satisfaction" that will "enlighten all men" and thus help society "to fulfil its obligation towards intellectual self-improvement."[21] At the opposite pole, Alvin Weinberg worried that the community of scientists might drift away from genuine public needs, pursuing the equivalent of "*l'art pour l'art.*" In a thoughtful response, Harvey Brooks attempted to find common ground among perspectives that seem incompatible.[22] In the end, however, the entire discussion was inconclusive, and it's hard not to sympathize with the pessimistic judgment of C. F. Carter:

> It seems to me that, though this is not entirely an economic matter, it is only from economics that any guidance will at present be obtained. In other words, it is possible to give some sort of answer to the question: What kind of distribution of scientific effort will most effectively increase the flow of wealth? But I see no means of finding out what distribution of effort will maximize human happiness, or maximize the joys of discovery.[23]

I've been attempting to go beyond Carter's pessimism. Earlier chapters have scrutinized (and sometimes rejected) categories the scholars of the 1960s took for granted, and, in formulating the ideal of well-ordered science, I've tried to answer the question Shils posed. In appraising major ventures in science policy, I've also endeavored to show how important that question is. Of course, the discussants in the debates of the 1960s might well regard my answer as defective—and their current descendants might legitimately complain that significant work needs to be done to connect the abstract ideal of well-ordered science with the kinds of decisions policymakers face. But I hope the ideal will serve as a first shot at the kind of standard we need, and will provoke others to refine (or replace) it and to do the empirical work of connecting it with the concrete decisions that now confront us.

Rather than trying to refine the ideal of well-ordered science further, my most important task at this stage is to take up an important objection, one that already surfaced at the end of chapter 10. Shouldn't the sciences challenge us and transform our values? Is well-ordered science altogether too comfortable?

21. Shils, ed., *Criteria for Scientific Development,* 19.

22. Harvey Brooks, *The Government of Science* (Cambridge: MIT Press, 1968), chap. 3.

23. Shils, ed., *Criteria for Scientific Development,* 37–38.

TWELVE

Subversive Truth and Ideals of Progress

THE GREAT ADVOCATES OF THE ENLIGHTENMENT, from the eighteenth century to the present, believed that by eradicating traditional prejudices and superstitions, scientific inquiry would enable people to live lives of superior quality. Behind this confidence is a far older view, expressed in Stoic texts extolling the benefits of knowledge and encapsulated in a famous verse from the New Testament: "And ye shall know the truth, and the truth shall make you free."[1] The evangelist was, of course, thinking about a very particular truth (or kind of truth) but the slogan has been co-opted by champions of reason who have no interest in a theological foundation for it—indeed who think that one of the ways in which the truth proves liberating is through exposing the errors and pretensions of religion. In a further irony, much of the rhetoric about the importance of seeking the truth seems to develop its own form of theology, viewing the high priests of the sciences as dedicated to a sacred task.[2]

Increasing our freedom is one way in which coming to recognize the truth might be good for us. I'll be concerned with a broader question—is knowledge always beneficial in some way or other? A tacit acceptance of affirmative answers typically underlies the claim that inquiry should proceed without stringent con-

1. John 8:32. The verse is quoted by Isaiah Berlin in "'From Hope and Fear Set Free,'" in *The Proper Study of Mankind* (New York: Farrar Straus Giroux, 1998).

2. This is evident in many of the writings and popular lectures of T. H. Huxley. In our own time, Carl Sagan and E. O. Wilson have continued the theme. It's worth noting that the idea of a "scientific priesthood" was not coined by critics but suggested by one of the members of the late Victorian circle of advocates of science, Darwin's cousin, Francis Galton. In *English Men of Science* (New York: Appleton, 1875), 195, Galton suggested that a "scientific priesthood" might guide the intellectual and moral life of the nation.

straints. Parodying the rhetoric of the theologians of inquiry, we might say that mundane concerns cannot be allowed to interfere with the higher work of those who serve in the temples of science.

Nobody who defends the value of knowledge can overlook the fact that discovering the truth sometimes diminishes human happiness. Patients who emerge white-faced from their doctors' consulting rooms, having learned that supposedly routine tests have disclosed dire conditions, find little cause for rejoicing in their enlightenment. Yet whether or not knowledge makes us happier, it might be better for us to have a clear view of the facts. The point has never been more eloquently—or more poignantly—expressed than by Thomas Henry Huxley, writing just after the sudden death of his beloved first child. Wracked with grief, Huxley wrote: "I could have fancied a devil scoffing at me . . . and asking me what profit it was to have stripped myself of the hopes and consolations of the mass of mankind? To which my only reply was & is Oh devil! truth is better than much profit."[3] The passage occurs in a letter to Charles Kingsley, the socialist clergyman and author, who had tried to comfort Huxley with the thought that he would be reunited with his son "in the hereafter." Huxley refused the consolation, on the grounds that he had dedicated himself to seeking the truth and that his search had disclosed the nonexistence of any resurrected state, leaving him bereft of the hopes of "the mass of mankind."

Previous chapters have shown, I hope, that no simple version of the thesis that it is always better for us to know the truth will do. There are vast oceans of truth that aren't worth exploring, and so the thesis must give way to the more plausible claim that it is always better for us to know *significant* truths. If there is no context-independent notion of significance, and epistemic significance is intertwined with past and present practical projects, then we cannot set the value of apprehending significant truths on some "higher" plane, so that inquiry must inevitably take precedence over everyday concerns. We've seen further that *applications* of the thesis that knowledge is good for us are problematic on two grounds, partly because our confidence in our capacity to obtain certain kinds of knowledge may be inflated, and partly because the impact of new knowledge on different groups of people may be radically different. Finally, in articulating the ideal of well-ordered science, I've envisaged one way in which a broader set of interests might be represented in the shaping of inquiry and the application of its results, but my proposal explicitly repudiated the idea that there's some higher standard that supersedes the preferences that would be formed in a process of ideal deliberation.

It's not hard to construct examples in which knowledge, whether acquired by an individual or flowing freely through the society, is far from beneficial. Recall

3. Leonard Huxley, ed., *Life and Letters of Thomas H. Huxley* (London: Macmillan, 1900), vol. 1, 233.

the case of the imaginary discovery that vast quantities of energy could be released by mixing readily obtainable ingredients in just the right proportions, a discovery whose widespread publication would make our world an extraordinarily risky place. Or, to extend the example of two paragraphs back, patients sometimes reasonably refuse diagnostic tests on the grounds that they know that they could not handle the bad news; to cite one dramatic example, less than 20% of the people at risk for Huntington's disease take the test to determine their condition, recognizing that there is no preventative strategy to be pursued and that any gains in freedom from being able to set their lives in order would be offset by the distress (or the psychological devastation) they would suffer in knowing in advance that their lives would end with the hideous decade-long degeneration they have observed in their close relatives. These are not just examples in which knowledge fails to increase happiness, but cases in which there's no discernible overall benefit.

Yet there's surely a genuine insight that drives the defenders of the thesis, even something noble about Huxley's avowal of the importance of truth. Start with the scenarios rehearsed in the last paragraph. Here it's natural to find fault with conditions that surround the acquisition of knowledge. Flow of information about explosive properties of everyday mixtures is dangerous because of the pathologies of individuals within our society (and maybe of society as a whole). In a just world inhabited only by the pure of heart, the knowledge would pose no threat. Similarly, if the majority of those at risk for Huntington's were as strong as the small group whose members do elect to take the test—or as strong as Huxley—they could genuinely benefit from knowing their status. The thesis that knowledge is good for us only founders on examples of these kinds, then, because we and the societies we construct are imperfect, morally flawed, deficient in character, unjust. As we aspire to rid our world and ourselves of these failings, a dedication to disclose the truth should figure as part of our overall ideal.

Even if we can't endorse the simple slogan, serious questions about the adequacy of my notion of well-ordered science as a normative standard remain. Inspired by Huxley, we recognize the possible value of subversive truth, knowledge that challenges the presuppositions on which the preferences held throughout a society are founded. Should we accept the traditional idea that recognizing the truth, even when it causes us pain, is, or at least ought to be, beneficial for us? And would acceptance of that idea require us to make different demands of inquiry than the promotion of collective well-being, as understood in the conception of well-ordered science? These are the questions I'll try to address.

Although many of those who descant on the virtues of scientific inquiry regard the dissemination of results that explode popular beliefs as having a bracing effect—much in the spirit, I sometimes think, of those Victorian educators who celebrated punishing exercise and cold baths for the growing boy—other voices sound very different themes. In the past centuries, various writers have complained about the world bequeathed to us by the advance of science. I'll focus on one example that has been, for many people, emblematic of the conflict between the sciences and our conceptions of what is valuable, the "war between science and religion" that came to a head in the mid-nineteenth century.

To understand the impact of nineteenth-century science on religious beliefs and, more generally, on the sense of what is valuable, we can follow the careers of any number of eminent Victorians. The case of George John Romanes will serve us well. In the 1870s, Romanes went up to Cambridge with the intention of entering the church, but he found his way into the study of natural science. After publishing a short book on the efficacy of prayer, in which he defended the possibility that God might answer human prayers without breaching the laws of nature, Romanes became more committed to the pursuit of science and joined the circle of thinkers around Darwin. Immersion in the life sciences apparently caused him to lose his faith, and at the conclusion of his second book, *A Candid Examination of Theism*, he declared both his inability to believe in the tenets of religion and the sense of loss that this produced. For Romanes, the feeling that his life had been diminished by the new understanding of nature was hardly a passing phase. During the ensuing years, he engaged in a protracted struggle to find value and meaning. A letter to Huxley underscores the dilemma that he—and many others—felt:

> Of course I have no doubt that in the long run Truth, however hideous, must prevail & I have little doubt that as it does so mankind will make the best of it, whatever it may be; but these considerations do not touch the question as to how far it is desirable that painful truths (or probabilities) should be rendered more painful by their rapid publication.[4]

Eventually Romanes rebelled against the claims of Huxley and his friends that a scientific—"naturalistic"—understanding of nature would cover all the phenomena, exploring a poetic vision that abstracted from the details of any orthodox theology but that repudiated a materialist approach to mind. Although he disavowed to the end many of the distinctive doctrines of Christianity, Romanes finally collapsed into the broad embrace of the liberal Anglican church.

Romanes' anguish was generated by his acceptance of two theses: first that the evidence of the life and earth sciences refutes the doctrines of Christianity (and other major world religions), and second that without the consolations of

4. The letter is dated 1879, one year after the publication of the *Candid Examination* (Boston: Houghton, Osgood and Co., 1878), and is quoted in Frank Turner, *Between Science and Religion* (New Haven: Yale University Press, 1974), 146.

religious belief life is shrunken and meaningless. He was hardly alone.[5] Yet it's important to recognize that the details of the dilemma varied from individual to individual, that which particular pieces of science were taken to be disturbing and which hopes were destroyed were hardly uniform. Late nineteenth-century judgments that the sciences had undermined religion rested in part on Hume's revival of ancient arguments, in part on the development of the earth sciences in the decades before Darwin, in part on the scientific work of Darwin and Huxley, in part on the "higher criticism" that emerged in German theological scholarship, in part on the growing awareness of the diversity of religious beliefs worldwide and of their functions in those societies that seemed most closely to preserve ancestral human conditions.

It may appear very easy to respond to concerns about the impact of the sciences on human projects and aspirations. People should be able to absorb subversive truths, and even benefit from them. But suggestions that complainers stop moping and join the party invite the rejoinder that jaunty confidence simply reflects the blunted sensibilities of those who do not fully appreciate what has been lost. Perhaps we can convince ourselves that we have come to terms with the destabilizing discoveries of the past (although the continued existence of people who object to educators' confronting children with those discoveries might give us pause), but that's only because we're unimaginative. To obtain a clearer vision of the complaint, we ought to imagine challenges to some presupposition of contemporary ways of attributing value to our endeavors or making sense of our lives. Suppose we were to find out that any of the following four propositions is true:

> There are genetic differences among groups that have traditionally been distinguished from one another, differences that fix limits to the ability of some people who have been targets of discrimination and that fix properties of temperament in those who have been dominant; there are no possible ways of compensating for these differences.
>
> Apparent human propensities to love and care for others are actually based upon manipulative strategies that are typically unconscious, and this is true of the most "concerned" and "altruistic" as well as those who are obviously exploitative.
>
> Stable human relationships are only possible in situations where people believe elaborate myths about themselves and their place in nature.

5. Turner's fascinating book *Between Science and Religion* studies six late Victorian thinkers who faced the same dilemma. For other useful perspectives, see James Moore, "Theodicy and Society: The Crisis of the Intelligentsia," and George Levine, "Scientific Discourse as an Alternative to Faith," both in *Victorian Faith in Crisis*, ed. Richard J. Helmstadter and Bernard Lightman (Stanford: Stanford University Press, 1990). Levine shows clearly how even the principal proponents of naturalism, like Huxley, solved the dilemma by essentially creating a theology of science.

Human choices, decisions, and actions are simple functions of physiological factors, many of them thoroughly banal and directly responsive to external causes (diet, regimes of exercise, and so forth).

Although it might be difficult to conceive how we could come to know any of these things, that isn't the point at issue. Those who urge breezy dismissals of complaints about science are asked simply to imagine that we gather overwhelming evidence for any one of them, and to reflect on the impact on our lives. The potential findings I've listed are sufficiently disconcerting to rebut the strategy of claiming that the pains are exaggerated. If there's to be a blanket defense of the idea that subversive truth is always good for us, it's going to have to depend on an *argument* for the superiority of knowing the truth, whatever it turns out to be.

Any such argument will have to take a stand on what makes things good or bad for us. It will be useful to begin with an abstract characterization of the familiar notion that people have plans for their lives and a tacit understanding of what matters to them, what gives their lives shape and significance. Of course, many human beings do not go through any explicit process of reflecting on possible goals and weighing their merits—although some do. Nevertheless, people's decisions and actions embody a conception of what is important, and, if pressed, they could articulate this conception, at least to some degree. A person's ideas about what is valuable may not admit of any straightforward ordering, in that there may be a plurality of things conceived as good and worth pursuing which are not easily fitted to any single scale. Even in such cases, however, the person may make decisions about how some goals are to be balanced against others. To a first approximation, how well the person's life goes depends on the character of these goals and the extent to which the person is able to attain them (in an appropriate balance); we'll consider particular ways of trying to give substance to this schematic account later.

Besides the goals people set for themselves, they have strategies for pursuing them. Hence we can regard a person's life as characterized by the *scheme of values*, consisting of the set of goals, the priorities among them, and the understanding of how the goals are to be balanced, and the *available strategies*, comprising the recognized means for achieving those goals. If the sciences and technology are functioning well for the person, then they'll supply means for expanding the available strategies, possibly even providing a way of promoting an end previously viewed as unattainable. Further, if one of the ends that's valued is satisfaction of curiosity about a particular topic (or cluster of topics), then scientific inquiry can contribute directly to the attainment of goals. But we can also envisage a number of subversive possibilities. First are examples in which, before the discovery, people have valued a particular goal and devised a single strategy for attaining it, only to learn that a key presupposition of this strategy is incorrect. Second are instances in which the discovery is inconsistent with claims that are essential to prior justifications of the value of the goal. Third are cases in which the discovery reveals that the valued goal is unattainable. Last are examples in which the discovery is taken to show that there is no basis for valuing the goal (perhaps because of the disclosure of properties inconsistent with the assignment of special value). Faced with the first two subversive discoveries, people may struggle to find alternative means of pursuing the goal or alternative justifications of its value. The last two are more challenging because they force people to abandon projects that have been important to them.

We can now formulate a general concern about subversive truths. Imagine a community in which inquiry discloses a new result that undermines prevailing conceptions of value in one of the four ways just mentioned. More exactly, at the initial stage, before the discovery, a large proportion of the community—perhaps even all members—have a scheme of values that recognizes a particular goal and a nonempty set of strategies for pursuing that goal. After the discovery, they have lost their previously perceived means of trying to attain the goal or they have lost their rationale for setting that as a goal for themselves or they have been given strong evidence for thinking that the goal is unattainable or they have been led to believe that the goal has nothing like the value with which it has been credited. At this intermediate stage, the system consisting of their scheme of values and their available strategies has been disrupted, possibly quite fundamentally if they have been led to deny the value of goals that were once central to their lives, and they must cast about for substitutes. Eventually, the members of the community, or their descendants, may arrive at a new scheme of values coupled with a new set of available strategies, absorbing the information that was once so disconcerting.

This schematic account reveals how any general argument for the beneficial character of subversive knowledge must proceed. To insist the truth is good for us is to contend that my three-stage process of the transformation of values constitutes a kind of progress. At the last stage, the community, or its successor, has arrived at a combination of a scheme of values and a set of available strategies superior to that with which it began. If a view like this is to be defended, two problems must be addressed. The first of these, the problem of short-term losses, notes that, from the perspective of those whose lives overlap the intermediate phase—particularly those whose lives end during this phase—the impact is severe. Such people live a significant portion of their lives, possibly the latter part of their lives or even their whole lives, without any way of structuring their projects and aspirations with which they can reflectively identify. They are constantly haunted by the suspicion that the things they do have no value, and, in this respect, they appear less well off than their unenlightened predecessors whose efforts to reach illusory goals may sometimes have led them to pursue, and even attain, other ends of genuine value. One way of looking at the people who occupy this unfortunate position is to see them as sacrificed to the march of inquiry. If we concede that their descendants, who arrive at a new scheme of values, are genuinely better off than the benighted ancestors of the first stage, then the intermediate (transitional) group suffers as part of a process that eventually promotes human well-being.

The second, and deeper, problem challenges the idea of progressiveness in values on which the general defense of the benefits of knowledge depends. In what sense, exactly, is it better to have replaced the scheme of values and set of available strategies in force at the first stage by that accepted at the finale? It's important to see that the simplest answers beg the question. If, for example, one notes that the original community is trapped by false beliefs, setting itself illusory goals and/or pursuing strategies that are doomed to failure, while its descendant has liberated itself from at least some of the errors, it still remains open why that matters. To appreciate the point, we can imagine a counterargument claiming that members of the ancestral group feel as happy as their successors—maybe even happier. Just as we think that these feelings of happiness don't translate directly into conclusions about which community is better off, so too the mere characterization of the earlier people as "trapped by illusion" doesn't clinch the case for progress. Having correct beliefs surely has something to do with the flourishing of a person's life, but so too does feeling happy. Until we attain a more substantial conception of what makes lives go well, the issue of how to understand the pertinent sort of progressiveness remains unresolved.

I'll deal rather briefly with the problem of short-term losses. Let's assume it's possible to provide an account of value progressivism vindicating the claim that, as inquiry advances, people come to adopt schemes of values and sets of available strategies superior to those they had had earlier. If this advancement were achieved by designating some particular subgroup whose members would be the necessary sufferers in the cause of value progress, then there would indeed be a familiar problem: it's simply unjust that the few should be plunged into misery for the benefit of the majority. The actual situation is not, however, quite like this. Given our assumption, the history of inquiry is one in which communities of people experience a trend of progress in values, although, at particular stages, some groups suffer the discomforts of the transitions. No particular collection of people is marked out for this unfortunate role, but, because the course of inquiry is unpredictable, it's just a matter of luck that some find themselves in an unenviable position in history (like Romanes for example).

A different analogy thus suggests itself. Collectively, societies often allow for risks of severe damage to a few so as to promote much tinier and less significant advantages to huge numbers of people. Consider the policy of setting speed limits. Were the limits on major roads to be set somewhat lower than they are, the frequency of fatalities and serious injuries would decrease—in other words, some people's lives would go far better than they actually do. There would be costs: delays, lowered productivity, perhaps a decrease in the range of products that consumers were able to buy. If we compare fortunate and unfortunate people under both regimes, and try to balance the losses incurred by those who would have emerged unscathed from driving under higher limits against the benefits that an unlucky person gains from lowered limits, then it's clear that the former are trivial compared to the latter. But, of course, there are vastly more people whose lives are not unalterably damaged or truncated by car accidents than people who suffer death or terrible injury in this way. On balance, we judge that the vast number of small losses from lowering the limits would be a greater burden than the relatively tiny number of huge gains. It's important to the legitimacy of our policy that there's no specifically marked group who will suffer: all of us have a minute probability of bad luck.

Let's now turn to the harder issue of understanding value-progressiveness.

What exactly could be meant by supposing that the transition from one scheme of values and set of available strategies to another is progressive? I'll consider four possibilities. In each instance, the guiding idea will be that those who stand at the end of the process of transforming values are likely to live lives of higher quality than those who come at the beginning. So we should start from the notion of the quality of life.

Historically, there have been many different conceptions of what makes lives go well. Without making controversial choices, it's not hard to recognize a number of elements that are balanced in the various proposals. Most obvious, of course, are the painful and pleasurable experiences of the person; hedonists famously propose to measure the quality of a human life through its attainment of pleasures and its avoidance of pain. Faced with apparent counterexamples, people who achieve their goals despite intense suffering, hedonists are forced to suppose that the fulfilment of the aspirations brings a higher and more exquisite pleasure, but it appears more plausible to broaden the scope of what's counted as making lives go well and to treat the attainment of ends separately and systematically. As I've suggested, people typically have plans for their lives, these plans mark out certain ends and desires as central, and one of the determinants of the quality of a human life is the extent to which central desires are satisfied, central ends attained. Again, it's possible to propose a pure theory of this general type, one that discounts pleasure and pain entirely, emphasizing only the achievement of central goals. Plainly, however, pure hedonism and pure goal-attainment are not the only options. One can propose a mix, maintaining that, other things being equal, lives go better if they contain more pleasure (or less pain) and, other things being equal, lives go better for those who achieve more of their central ends, and one can suggest ways of trading desire-satisfaction against hedonic accumulation.

Two other factors need to be considered. Some writers have felt that it's important to the quality of a human life that its plan be the authentic expression of the person who chooses it. Aware of the ease with which our preferences can be imposed from without—from our cultural tradition and the society in which we grow up—they suggest the character of the decision process through which a "life project" is shaped affects the quality of that life. If we suppose that freedom, or autonomous choice, is a matter of degree, that people can be more or less coerced by parents, teachers, and fellow citizens, then we can now consider a family of three-dimensional theories: along the first dimension the quality of life is assessed by the degree to which people are free in forming their central desires and aspirations, along the second it is measured by the extent to which those central desires and aspirations are satisfied, and along the third it is evaluated by the balance of pleasure and pain. For theories of this general sort there are straightforward intradimensional orderings—more freedom is better, more satisfaction of desires is better, more pleasure is better—and the accounts are distinguished by their interdimensional trading.

The last factor to be considered is the intrinsic character of the project that is chosen. Quite independently of whether the project is the result of free deliberation, we can ask whether the life goals a person has set are really worthwhile. It seems eminently possible that a person should freely choose a life centered on quite trivial goals, or, alternatively, be pressured into a life course comprising aspirations that were genuinely important. Postponing, for the time being, the way in which judgments about the intrinsic character of goals might be grounded, let's expand our family of accounts to embrace four dimensions, adding the intrinsic character of the goals. In this last case, the intradimensional scale is obviously much more obscure, and we may await proposals from theorists of different kinds. It's worth noting, however, that we have now made space for one of the most prominent historical views about the quality of human lives, the idea that there is a set of privileged tasks set for members of our species by a personal deity and that how well our lives go has everything to do with our success at these tasks and (for example) nothing at all to do with our pleasures and pains.

Now that we have an inclusive framework within which views of the quality of human lives can be set, it might seem that the next step would be to evaluate ways of balancing the different dimensions and specifying the ordering along the "intrinsic character of goals" dimension, if that is to be given some weight. Any attempt to take this step would lead me away from the ecumenical approach I've adopted so far. In considering four ways of generating a notion of the progressive transformation of values, I'll proceed rather differently, in effect allowing each proposal to develop its preferred account within the given framework. For my ultimate goal is to understand the possibilities of an account of value progress that will underwrite a general argument for the benefits of knowing the truth, and this can be achieved if we review the options, fitting to each the account of the quality of life that best suits it.

Start with the simplest proposal, one that avoids any ranking of goals according to their "intrinsic character": there is nothing to choose among schemes of values except on grounds of consistency or attainability. The proposal then comes in two parts—first a claim that for any two consistent schemes of values neither is intrinsically superior to the other; second, the suggestion that the quality of lives is measured by the extent to which goals are attained, desires fulfilled. In effect, then, the account of the quality of life gives full weight to the dimension that represents satisfaction of central desires, none to considerations of autonomy, none to pleasure, and none to the "intrinsic character" of the desires. The last is plainly a consequence of the first part of the proposal. Omitting any reflection of the sum of pleasures is clearly advantageous for the defense of the benefits of acquiring knowledge, since, as we've seen, recognizing the truth can bring considerable anguish. On the other hand, one might suppose that the case for the value of truth could be enhanced by giving weight to the autonomy dimension, recapitulating the evangelist's theme that "the truth shall set you free." I'll explore this idea as the second of the options I review.

Now for the general argument that inquiry is likely to be conducive to a progressive transformation of values. Coming to know the truth can teach us any of a number of things: how to attain goals we'd previously set for ourselves, that certain strategies we'd pursued will not enable us to reach our ends, that some of our ends are unattainable, that valuing some of them is at odds with our other beliefs. If we respond to this knowledge by amending our scheme of values, we're likely to relinquish strategies and goals that frustrated our efforts in favor of projects that are more auspicious. Of course, it's not *certain* that we'll do better at satisfying our desires, but there are grounds for thinking that the *expectation* of fulfilment is greater.

The fundamental point is that replacing the earlier goals, with their attendant difficulties, is relatively easy. Precisely because there are no intrinsic differences in the character of our projects, surrogates for the old scheme are straightforwardly introduced. Since these are likely to remedy earlier inconsistencies or to specify ends that are attained more often, the proposal would count the transition as progressive because the quality of the lives of the descendant community is expected to be higher than that of the ancestral community. On the only dimension of quality of life that matters, desire satisfaction, enlightened latecomers score higher than their benighted predecessors.

Precisely because there are no constraints on the substitute schemes of values, this entire account and the argument it supports are too facile. Recall the angst-ridden Victorians, Romanes prominent among them, who struggled to find new aspirations in the light of scientific findings that shattered their earlier plans, goals, and sense of themselves—or imagine our own predicament if we discovered that some radical sociobiological hypothesis about the springs of human nature were correct. To suggest that the disruption of an older system of values involves no loss, because anywhere we look we can find a replacement, is, at best, to be grossly insensitive. It's hardly reassuring to be told that people can make up for the loss of a personal deity or the value they attach to human relations by dedicating themselves to collecting paper clips or improving the tone of their abdominal muscles. How badly this unconstrained desire-satisfaction account serves the purposes of defending the value of truth can be seen by reminding ourselves of Huxley's eloquent response to Kingsley—his point was that truth is *better than* much profit, not that it is *the key to* much profit (that is, to satisfying our desires, whatever they chance to be).

Let's now consider a second pure proposal, the claim that nothing matters except the autonomy of the person in deciding what goals to pursue. The argument for the progressiveness of the transformation of values will now take the fundamental criterion to consist in the extent to which those who come at the end of the process can be expected to reach more authentic decisions about how to structure their lives, since it's precisely this that will entail that those lives have higher quality. So the defense will rest on the premise that increased knowledge will promote human freedom. The long history of accepting this premise testifies to the plausibility of the idea that, when people are given a clearer view of the genuine options, they are able to identify more clearly what is of principal concern to them and what is worth striving for.

There are several difficulties with this line of reasoning. First, it is not evident that acquiring knowledge always increases human freedom. To perceive our own predicament more clearly sometimes causes us to refrain from endeavors that are doomed to frustration, but it need do no more than expose the failure of apparent possibilities. Our freedom would be augmented if at the same time new options were disclosed to us, and, while that is possible, it isn't inevitable.[6]

Second, it's not plausible to suppose that all the products of inquiry will have any bearing on the autonomy of decisions about which projects to pursue. Perhaps answers to questions taken to be practically significant will sometimes enable people to formulate plans that express their desires more adequately—as, for example, when the new findings reveal that ends earlier taken to be incompatible can be reconciled—but there are many types of epistemically significant inquiry that lack any such power. We saw in chapter 6 that such inquiries will

6. Isaiah Berlin makes the point eloquently: "To discover that I cannot do what I once believed that I could will render me more rational—I shall not beat my head against stone walls—but it will not necessarily make me freer; there may be stone walls wherever I look; I may myself be a portion of one; a stone myself, only dreaming of being free" ("'From Hope and Fear Set Free,'" 104; see also 108, 114–115).

satisfy natural curiosity, and supposed there that obtaining such satisfaction would be good for us, but, from the perspective of the line of argument now being considered, such satisfactions only contribute to the quality of lives insofar as they promote free decisions about central projects. To say the least, any such connection appears highly strained.

Third, increased knowledge sometimes hinders us from achieving other things that would be beneficial for us. As a matter of psychological fact, people who learn certain kinds of things about themselves or about the nonhuman world may find themselves unable to carry out activities that have formerly been important to their lives: from the mundane example of the golfer whose theoretical knowledge of the mechanics of the swing undermines his ability to hit the ball consistently to the artist or writer who only undertakes a massive project because of her initial ignorance about the labors it will involve. Knowledge of certain kinds can thus be expected to decrease freedom, through inhibiting people from expressing important facets of their characters.

Fourth, the appeal to freedom in choosing one's central projects as the ground of the quality of life is just as vulnerable as the earlier desire-satisfaction theory to the objection that the plans a person sets might focus on the trivial (or worse). It would be wrong to rate the quality of a life devoted to paper-clip-collecting or abdominal-toning as high on the grounds that it proceeded from a free decision to pursue these ends. Instead, we'd surely lament the fact that the person's interests, genuinely expressed in the decision, were so blind and banal.

Fifth, and finally, there's an obvious problem with the pure account's neglect of any consideration of the extent to which central aspirations are satisfied. If the entire focus is on the autonomy of the process in which the guiding desires are identified, then lives count as having equally high quality whether the desires are largely satisfied or universally frustrated. Even worse, a life in which the decisions are slightly freer than those in another, but in which the ends are never achieved, ranks as higher in quality than the other, even though virtually all of the second person's desires might be satisfied. Consequences like these are surely counterintuitive, and they tell against the complete emphasis on autonomy.

My catalogue of objections is intended to separate important issues. In response to the last, there's an obvious remedy: adopt an account that mixes the considerations of the first two options. The quality of lives is assessed along two dimensions, according to the extent to which the decisions about central projects are made freely and reflecting the degree to which central desires are satisfied. How are the scores along the two dimensions weighed to produce an overall measure? I shall not try to answer this question. Plainly some assignments of weights will answer my fifth objection, but the important point to observe is that the new hybrid theory doesn't touch the first four concerns. In particular, the fourth criticism, which raises the possibility of deciding to pursue trivial projects, applies not just to both pure proposals but also to the full range of mixtures. No matter how the dimensions are balanced against one another, the resultant ranking will give high marks to lives whose freely chosen—but intuitively worthless—ends are frequently attained.

Plainly, if the defender of the progressiveness of the transformation of values through recognizing truth is to advance a general argument it will be necessary to bring explicitly into the picture the "intrinsic character" of the projects that are set. We need a compelling account of the worth of certain kinds of projects, and an argument for thinking these projects are more likely to be taken as central or more likely to be fulfilled by those who stand at the end of the transition from one scheme of values to its successor. There are two possibilities, one that views value as generated by the decisions of an ideally situated agent, and one that regards value as independent of human beings and their deliberations.

I'll start with the first of these options. The guiding idea is that the intrinsic worth of a desire or a goal is measured by its proximity to the desires or goals that would be adopted by ideal agents who had uncoerced opportunities to set down their aims in light of ideal information. Lives achieve high quality in three ways: if their central goals are close to those that would be adopted by an ideally unconstrained agent in the ideal judgmental situation; if those goals are themselves freely chosen; and if a significant proportion of the goals are attained. How should we make trades among these factors? The process that transforms values is likely to generate lives of higher quality in the descendant community if increased knowledge raises the probability that the goals achieved will be closer to those that would have been adopted in the ideal situation, if recognizing the truth promotes freedom, and if the insights achieved enable people to satisfy more of their central desires. We've seen already that the second idea is problematic, and some of the problems discerned above continue to apply in the new situation (for example, the first three of the five difficulties). Hence, I'll suppose that the defense of the value of truth goes best if the autonomy dimension is assigned relatively low weight. If the greatest weight were given to the third dimension, effectively taking the satisfaction of desires to be most important, we'd revert to a position quite close to the first option I considered, and would be vulnerable to the possibility that a life in which trivial projects succeeded would be judged of higher quality than one that aspired to things of genuine value but in which the efforts often, although not always, were in vain. Since this possibility ought to be debarred, we should conclude that greatest weight ought to be assigned to the measurement of the intrinsic character of the central goals. Hence the argument will succeed or fail according to whether it's possible to make the case for claiming that knowledge gained through inquiry is likely to lead us to make judgments of value that accord more closely with those that would be generated from the ideal situation.

A relatively simple line of reasoning would seem to deliver just what's needed. As we gain knowledge through inquiry we come closer to the condition of full information which serves as the basis for making judgments of the worth of various endeavors. It doesn't follow from the fact that we know more that the judgments of value we make are *certain* to be more similar to those resulting from the ideal state than the assessments flowing from our earlier ignorance—for there's an analogue of the familiar point that acquiring more evidence can lead one further away from the truth. Nevertheless, relying on analogies with other epistemic predicaments, we can argue that greater knowledge would be *likely* to produce judgments that accord more closely with those made on the basis of the ideal maximum of information. So the schemes of values adopted in the advance of inquiry are likely to improve on their predecessors.

Unfortunately, this reasoning begs the crucial question. Recall the fundamental characterization of the intrinsic worth of ends: the truly worthwhile ends are those that would be chosen by an ideal agent in a state of ideal knowledge. The reasoning just offered tacitly elaborates this characterization by making two substantive assumptions. First, it's supposed that the state of ideal knowledge simply consists in obtaining the maximum set of true beliefs of the type generated through scientific inquiry. Second, it's taken for granted that, in acquiring these kinds of beliefs, people do not take on characteristics that interfere with their performance as ideal deliberators. Critics of the view that the truth is invariably good for us will object to one or both of the assumptions.

The first is vulnerable on two grounds. To equate the state of ideal knowledge with some imagined extension of the process of inquiry is to grant to the sciences a monopoly on the pertinent forms of cognition. One form of the critics' objection is that there are types of knowledge that aren't attained through empirical inquiry—think of the suggestion that poets and novelists gain insights into nature and humanity that are of quite a different kind from any attainable by scientific research. Defenders of the value of truth can try to block this response by suggesting that the states conjured in the criticism aren't genuinely states of knowledge, reserving the term 'knowledge' for beliefs rooted in the procedures of the sciences and the everyday investigations from which the sciences grow. But the more narrowly one defines the notion the less plausible is the claim that the ideal situation for making value judgments is a state of ideal *knowledge.* To see this, let's scrutinize the inference from claiming that value is that which is recognized in an ideal situation to the conclusion that value is that which is recognized under a condition of ideal knowledge. Critics are likely to protest that this ignores the refinement of emotional responses to others or aesthetic appreciation, both of which ought to figure in the assessment of value. Given the decision to construe knowledge narrowly, these have been exiled beyond the bounds of the epistemic, with the result that the conception of the ideal situation overlooks features that the critics regard as crucial. In effect, then, the argument confronts a dilemma: either it must suppose that all knowledge flows from the standard practices of the sciences, in which case the epistemic characterization of the ideal situation begs the question, or it can accommodate some of the critics' points about sensibility by broadening the conception of knowledge, thereby undercutting the idea that all the pertinent kinds of knowledge are obtained through scientific inquiry.

Turn now to the second major assumption. Part of what motivates worries about the value of subversive truths is the idea that acquiring certain types of knowledge can stifle other capacities. As we've already seen in the discussion of the view that the quality of life is determined by the autonomy of the decision of what goals are to be central, there are prima facie examples in which increased knowledge inhibits other psychological dispositions. To make a general case for the value of knowledge, it's thus necessary to argue that the ideal deliberation either doesn't require any particular capacities in the agent (such as the species of sensibility on which some opponents insist) or that there's a guarantee that such capacities will survive intact. Until these lacunae have been filled the defense fails.

The last attempt I'll consider is based on an objectivist approach to values. Objectivists claim that certain ends are worthy of our pursuit, whether or not we find them attractive. Insofar as they endorse the idea just canvassed that such ends would be recognized as valuable by an ideal deliberator in an ideal situation, they invert the perspective: those ends are not valuable because they would be singled out under those circumstances, rather what makes the situation and the deliberator ideal is that they combine to produce the correct judgments about what's independently valuable. As in the previous case, we must suppose that the dimension that contributes most to determining the quality of a life is the one that records the intrinsic character of the goals and desires that the person views as central (for otherwise we'd encounter the problems facing earlier efforts). Hence the crucial question is whether there's a general argument for claiming that coming to know the truth is likely to enable people to identify more clearly those things that are especially worthy of their aspirations.

Objectivism comes in varieties, and we'll explore some distinctions shortly, but one component of all objectivist accounts is a catalogue of what things are important for us. Let's dub this catalogue *the List.* Now one obvious way to support the value of knowledge is to offer a List that includes only the sorts of qualities that are readily and directly promoted by the advance of inquiry. So, for example, one might suggest that the primary item on the List is the goal of being a clear-headed and informed person. If that were a plausible version of the List, then it's easy to envisage how to develop an argument for the general value of greater knowledge.

Assume, then, that there are some Lists that specify goals for us that we're likely to perceive more clearly and to attain more readily if we gain greater knowledge, and for which this feature can be shown fairly directly. Call such Lists *narrow*. The trouble is that there are other candidate Lists which include, and even give highest priority to, goals that lack any direct connection with human knowledge. So it may be proposed that our central aims ought to be to develop close personal relationships or to refine our appreciation of the beauties of art and nature or to promote the stability and harmony of the community to which we belong. Lists which incorporate ends like these are *broad*.

Objectivists who want to defend the general value of inquiry can proceed in any of three ways: they can argue that the correct List is a narrow List (in which case they'll have a direct argument for their conclusion); they can argue that the correct List is a broad List, but that the acquisition of knowledge advances (or at least doesn't interfere with) the recognition and attainment of the nonepistemic ends that are on the List; or they can try to proceed without trying to settle issues of the correctness of rival Lists, providing reasons for thinking that greater knowledge is valuable from the perspective of any candidate List.

Unless it's conceived as a piecemeal strategy, one that surveys the contents of the Lists actually proposed, showing in each instance that the favored ends are either goals that are not worth endorsing or goals to which inquiry is at worst neutral, I think the last approach is hopeless, for a general argument would require some ability to bound the potential candidates. So defenders are forced to tackle the difficult task of singling out some List or cluster of Lists either as giving the truth about what's objectively worthwhile or at least revealing the neighborhood in which the truth lies.

How can they undertake that task? Appeals to the authority of sacred texts or to the deliverances of some "intuition of value" are hardly likely to prove convincing. I'll suppose, then, that the only promising approach is to identify some facts about people from which the contents of the correct List can be derived. What facts might these be? As I've suggested at several earlier stages of the discussion, adverting to human happiness to try to ground what's genuinely valuable for us to pursue is not likely to sustain a general argument for the benefits of knowledge. Precisely the problem is to explain why results that deprive us "of the consolations of the mass of mankind" are nevertheless to be welcomed. The most promising attempt to fill the gap is to appeal to the idea that particular kinds of ends are constitutive of our nature and to see human lives as going well if they identify, aim at, and achieve these ends.

Of course, human beings have plenty of properties. Which ones are to constitute our nature? If singling out the essential properties were simply a matter of philosophical intuition, then objectivists would have made no progress, for the proponents of alternative Lists would doubtless assert the power of their rival intuitions. We do better to take a clue from essentialist doctrines about physical and chemical substances. To the extent that it's plausible to hold that the essence of gold is to have atomic number 79 or that the essence of water is

H_2O, it's because the attribution of these properties figures in the explanatory accounts of why chunks of gold and quantities of water have the attributes that are common to gold and water, respectively. Physics and chemistry supply systematic ways of explaining the properties of substances in terms of atomic and molecular characteristics, and this pervasiveness of the atomic and molecular characteristics in explanations convinces us that these are the "deep" properties, the ones essential to the substances being what they are. In the same way, we might hope that biology or psychology might provide systematic explanations of the properties of human beings and that the properties that figure in these explanations (the analogues of atomic and molecular structure) would constitute our nature.

Under inspection the hope evaporates. To begin with, the atttributes all human beings share do not include all the "higher capacities" that inspire objectivists to construct their Lists. There are unfortunate members of our species who are unable to think or to feel any sympathy for others, for instance. Any use of the explanatory strategy for discerning essences that has a chance of serving the objectivist's turn will have to debar such individuals, viewing them as abnormal. A statistical notion of normality will not serve here—for it's a contingent fact (and one that might be subverted in the wake of a nuclear holocaust) that births typically do not give rise to human beings with these unfortunate deficiencies. The obvious thing to say is that these human beings are abnormal because they don't have the important—essential?—human characteristics, but this is to give up the enterprise of employing the explanatory strategy to discover those characteristics and to return us to the predicament of dogmatically trading Lists, the predicament from which objectivism was struggling to escape.

An appeal to our evolutionary history might seem to promise a way out. Perhaps we can identify functions in terms of the ways natural selection has shaped our species (or our hominid, or primate, ancestors) and so pick out those traits that it's important to develop? There are two general difficulties with this approach. First, as numerous thinkers have observed, there are many evolutionary legacies that we want to resist, limit, or overcome, rather than endorse or develop. Second, there are simply far too many ways in which our bodies (and brains) have been shaped by evolution.

The second point reveals how objectivism is bound within the circularity we've already seen. Those who try to tease our objective good out of our human essence are fond of seeing certain physical regimes as worth pursuing because they develop organs and systems whose functioning explains our activities. They celebrate athletic programs that enhance musculature, cardiovascular function, lung capacity and so forth, on the grounds that human movement is explained in terms of muscles fueled by heart and lungs (and, of course, they can gesture at evolutionary stories of why the hearts and lungs do what they do). Only a moment's reflection is needed to notice that there are all kinds of organs and systems whose development we *don't* regard as intrinsically valuable. Those who can eat vast quantities of food display the efficiency of their digestive systems, fraternity contests reveal differences in the ability to metabolize alcohol, and, with imagination, we can envisage similar manifestations of the fine development of sweat glands or the urinary system (each of these backed equally with an evolutionary tale). There's no reason to take the underlying physical properties as less essential to our species than the ones objectivists typically prefer to emphasize—unless, of course, one has already tacitly used prior judgments about what is valuable to pick out those aspects of our biology that are to be essential and that are to ground claims about what's intrinsically good for us.

More to the point, exactly the same trouble affects attempts to identify particular psychological traits and capacities as constitutive of our nature and therefore particularly worthy of development. One obvious way to defend the overriding importance of seeking the truth is to claim that a yearning to satisfy curiosity is essential to being human—"All men by nature desire to know." So one might contend that the entire spectrum of human actions can be explained in terms of rationality, that actions result from rational consideration of how best to achieve one's ends. It seems clear that any sense in which this thesis is plausible involves so attenuated a notion of rationality that it can hardly support the view that the process of inquiry should be seen as the development of our rational capacities. Furthermore, even if the notion of action is restricted to ensure that every action issues from a desire and a belief about how to satisfy that desire, there's an enormous amount of human behavior for which this type of explanation is, at best, controversial. The most consistently rational members of our species also breathe, blink, burp, flinch, dream, obsess, burst into song, become angry, fall in love, and so forth. Unless all these varieties of behavior proceed from underlying beliefs and desires, there's simply no grounds for thinking that everything we do is the expression of the rationality that is central to our nature. Once again, the argument makes a selection from the range of phenomena, claiming that some of them are important, that these are to be explained in terms of a capacity that is essential to us, and that development of this capacity is objectively valuable for us. Even before one probes the details of precisely what form of rationality is thus incorporated into our nature, we ought to be suspicious about the ways in which the phenomena are selected, seeing the elevation of intentional action over (say) shivering or daydreaming as already making just the assumptions about value that were supposed to be established as conclusions.

An objectivist defense of the benefits of knowledge can't succeed by invoking the idea of the human essence. In the end, I think, objectivists have little more to offer than the bare assertion of their favorite Lists, and, in consequence, this last option, like its three predecessors, fails to deliver the general argument we've been seeking.

I have attempted a systematic survey of all the possibilities for showing that "truth is better than much profit" and have come up empty. Indeed, what is most striking is the fact that, when articulated, the possible lines of defense are so unpromising. From the Enlightenment on, those committed to the value of scientific inquiry have been so convinced that challenges could be quickly dismissed that they have failed to explore the bases of their conviction. Behind the often evangelical rhetoric about the value of knowledge stands a serious theology, an unexamined faith that pursuing inquiry will be good for us, even when it transforms our schemes of values. It's time to abandon that theology too. We need agnosticism all the way down.

THIRTEEN

The Luddites' Laments

A NINETEENTH-CENTURY PROTEST against industrialization has lent its name to the general impulse to protect the impact of science on society. Luddites do not necessarily attack machines. They may complain that the sciences have deprived us of faith in a deity, or that they have estranged us from the beauties of the natural world, or that they have raped nature and overridden the values of women and people of color, or that they have created technological societies in which human lives are reduced to a single dimension. Creationists, Romantics, feminists, postcolonialists, and critical theorists can all march behind the Luddite banner.

If there's no general argument for the claim that greater knowledge improves the quality of human lives, then maybe the Luddites' laments are justified. A moment's reflection on the argument of the last chapter suggests there will be no general defense of the complaint: the thesis that knowledge is always bad for us is far less plausible than its contrary. So the issue turns on the merits of particular cases. Are there instances of scientific knowledge that have worsened the human predicament?

I'll look briefly at two examples, the scientific subversion of religious doctrines and the rise of the technological society. Attention to these cases will, I believe, reveal how other versions of the lament should be evaluated. As we'll see, the Luddite critiques raise genuine issues, but, in the end, they reinforce my claim for the normative ideal of well-ordered science.

Luddites suppose the acquisition of particular truths inevitably diminishes the quality of human lives; it's not just that those truths were mishandled, so that the blame attaches to those who tried to apply them; rather the discovery of the truths leaves no possibility for benign use. More precisely, once members of a society come to believe those truths then there is no way for them to conduct their lives and retain the quality of life their ignorant predecessors enjoyed. Whose lives? The critical case against science can be made more or less ambitiously. On the stronger version, the complaint is that a discovery makes everyone's life go less well. The weaker version concedes that the discovery would not preclude some lives attaining equally high quality, despite its overall impact being negative.

Although the point is controversial, many people (for example, the despairing Romanes) have believed that an accumulation of claims in geology, biology, anthropology, history, and sociology is inconsistent with the doctrine that there is a personal deity who is concerned with our well-being.[1] Suppose, for the sake of argument, the incompatibility is real. Recognition of the truth of the pertinent pieces of science then generates the first Luddite complaint I want to examine. Let's start with the strong version, the idea that debunking theism has diminished the quality of all human lives.

How might this be defended? Here are three possible strategies, corresponding to three of the four dimensions that figured in the analyses of quality of lives in the previous chapter. The simplest proposes that people who didn't know the results from the sciences were happier, other things being equal, because they could retain hopes of a benign deity and of a future state in which they would be permanently reunited with their loved ones (they are better off along the hedonic dimension). A second idea takes the effect of loss of faith to decrease the ability to attain ends that might have been achieved, such as the cultivation of loving relationships with others (a loss in terms of the satisfaction of central desires). A third views those without faith as unable to formulate projects for their lives that are as intrinsically worthwhile as those available to theists. (The fourth dimension—loss of autonomy—is not represented because I've been unable to think of a plausible way in which the scientific discoveries might diminish our free choice of projects.)

Consider the second Luddite strategy. The person who has heard the terrible news from science is supposed to be less able to achieve the most important ends. These ends can't be any state of union with the deity, since it's a presupposition of the debate that the discoveries show that there is no such being. Instead, to make the strategy work, Luddites must maintain that, as a general mat-

1. The inconsistency is defended both by nonreligious people who view theism as decisively refuted and by some religious believers who take it to show the inadequacy of science. Liberal theologians and some humanists have held that the scientific findings are compatible with belief in a personal deity. In the final chapter of *Abusing Science* (Cambridge: MIT Press, 1982), Patricia Kitcher and I tried to defend their compatibility. It now seems to me that our plea for peaceful coexistence (for which I must take primary responsibility) was too facile, and that I should have been persuaded by the more clear-headed arguments of my co-author.

ter of human psychology, other goals, genuinely possible, are inevitably less thoroughly pursued and less completely attained after the loss of faith. Taking the achievement of loving relationships as exemplary, the argument presupposes that it's a psychological generalization that those deprived of faith can't achieve loving relationships that compare with those of the devout. That generalization is decisively refuted by the many examples of love and friendship among nonreligious people.

The point extends to a partial rejoinder to the third strategy. If Luddites concede there are some goals worthy of pursuit independently of the existence of a deity, then they must either suppose that schemes of values that posit those ends and conjoin them with ends that are deity-dependent are superior to schemes of values that just embody the deity-independent ends, or they must submit that, as a psychological generalization, identification of the appropriate ends requires belief in the deity. The former idea is implausible, since there's no discernible reason why valuing love and friendship (say) should be inferior to a state in which one's central ends include both love and friendship and entering into community with a deity, given that the latter goal is, thanks to its false presupposition, unattainable. The appeal to psychological generalization does no better, however, since, as before, we can advert to secular lives that celebrate love and friendship as counterexamples to the alleged regularity.

All this has assumed that there's some residual set of deity-independent values that are genuinely worthwhile. Luddites might, however, articulate a version of nihilism: if there is no God, nothing is genuinely worth pursuing. By assumption, the scientific discoveries reveal the nonexistence of the deity and our enlightened predicament is therefore one in which there are no intrinsically worthwhile goals for us. One way to respond would be to point to putatively valuable human ends, such as the relationships that have served as my examples. But there's a deeper point. How exactly has the discovery worsened the human predicament? Along the dimension that measures quality of life by the intrinsic character of the projects pursued, people can be no worse off than before, since they have gone from a state of pursuing goals of no importance to a state in which they know that there are no goals that have any importance. All that has been lost is an illusion, and, on pain of contradiction, Luddites can't argue that the sole worthwhile human project is to maintain this illusion. So the only way the complaint can succeed is by looking at other determinants of the quality of lives, either holding that the illusion is needed to attain whatever goals we now set for ourselves (a position which we've already found wanting, and one that's doubly problematic now it's recognized that those goals are not worthwhile) or else that the illusion increases pleasure and diminishes pain, a view to which I now turn.

On the face of it, the first of my three suggested strategies, which emphasizes hedonic contributions to the quality of life, seems a rather peculiar way to insist on the diminution of human well-being caused by the complex of discoveries that undermine faith. Yet if critics want to make their case on the basis of psychological generalizations about the consequences of forsaking belief in a providential deity, then they are on safer ground in appealing to considerations of comfort and anguish than in venturing the speculations that have figured in earlier phases of the discussion. For it's much more plausible to suppose that people would be happier if they could believe in their eventual union with the deity and in their reunion with loved ones in a future state of bliss: Huxley was perceptive in characterizing his loss as that of "the consolations of the mass of mankind." Human lives vary in a large number of respects, some of them being subject to far greater strains than others, but, critics may reasonably contend, some suffering is inevitable and, in particular, virtually everyone has to confront the death of a much-loved person. At such moments of stress, the pains would be softened by the thought of a benign otherworldly future, illusory though that is. Hence, for any postenlightenment life, whether it be one full of challenge, pain, and sadness or one that has only occasional moments of grief, a pre-enlightenment counterpart would be happier. (Call this the *no-atheists-in-foxholes* argument, or NAIF for short.)

NAIF can be supported by reading deeply in the writings of the great late Victorian Doubters, especially those, like Romanes, who agonized over a period of years. The anguish may seem so intense as to support the judgment that *this individual* would have been better off without the disturbing knowledge. But, of course, the variation in unhappiness is enormous. Only a naif would suppose that because everyone is afflicted to some degree and because some are afflicted to a large degree, all human lives have suffered a severe diminution in quality because of the eighteenth- and nineteenth-century discoveries. The critics' psychological generalization must be qualified through recognizing that contemporary secularists have developed psychological mechanisms for coping with adversity without turning to a nonexistent deity. NAIF has to concede that, in at least some instances, the differences in level of pain felt by those who can take illusory comfort and those who reject such comfort are not great.

In the last chapter I imagined someone suspicious of the benefits of knowledge insisting that the contemporary sense of having come to terms with the unsettling discoveries of the past reflects the fact that our sensibilities have been blunted. Is the response of the last paragraph vulnerable to the same complaint? No. If the issue was that the lowering of the quality of lives through people's decreased ability to identify genuinely worthwhile central ends, the envisaged objection would be pertinent. I've already examined the debate on those terms and replied to the critics' charges. What now concerns us is the objection that life's hedonic tone is the principal factor in measuring quality, and, in this context, it hardly matters that happiness is achieved and anguish avoided by dulling our sensitivity to a particular kind of pain. All that matters is that the ways of coping be effective.

Nonetheless NAIF still finds some difference between lives before and after the scientific advances, at hedonic gain to the former. Hence the crucial case for evaluating the argument is to imagine a person who has thoroughly absorbed the unsettling discoveries and who has well-developed methods of coping with the reversals of life. NAIF is committed to supposing that this person's life would have been improved, other things equal, by subtracting the knowledge and adding the lost faith. But if the methods for coping are genuinely well developed, as they sometimes are, then the gains in terms of diminished pain will be small, while the costs of diverting attention from attainable ends (relationships with others, say) to unattainable ends (propitiating the deity) may be more considerable. So NAIF has to give great weight to pleasure and absence of pain. An obvious consequence of this approach to the quality of life will be to elevate the levels of well-being of people with disrupted development, who are kept in a state of gentle bliss by regular injections of drugs. The life of a satisfied pig will be ranked ahead of that of dissatisfied Socrates.

Although I've been examining the strong version of the Luddite complaint —all our lives are diminished by the discoveries that debunk theism—many of the points apply to the less ambitious version as well. If the weaker charge that the discoveries inevitably have an adverse impact on the lives of a large number of people is to succeed where the universal claim fails, then it will have to be shown how the compensatory strategies that enable some to avoid lowered levels of well-being can't be extended to everyone (or almost everyone). The most plausible way of doing this would be to extend the line of Luddite argument just considered. So it might be conceded that there are a few individuals for whom the increased pain that comes with loss of faith is rather small, and hence compensated for by their clearer vision of what to pursue, even though this cannot be so for "the mass of mankind." That would be to advance a quite unsubstantiated conjecture to the effect that the capacity for coping with reversals of fortune is a matter of the intrinsic strength (or thickness of skin?) a person has. An alternative hypothesis would propose that the character of the social environment, perhaps most importantly of early education, is the most important determinant of someone's ability to face adversity without the consolations of faith.

It is an unhappy truth that some lives are afflicted by severe misery and stress. Equally, it has to be admitted that many people have found the prospects dangled before them by various religions to be sources of comfort, so that the scientific findings that seem to belie religious doctrines are perceived as threatening, even potentially devastating. Very probably, a significant amount of human misery could be relieved and, in some instances, religious traditions themselves surely serve as obstacles to the relief. More important, however, for present purposes, the early offer of a religious crutch may interfere with the development of other ways of responding to adversity. It may well be possible to devise systems of education that make the benefits of scientific knowledge available to all and that equip children to cope with personal misfortune, even if they lack comforting illusions about a divine providence.

The pains felt by Romanes and by many like him are real, and they should not be denied. It doesn't follow that the only way to avoid thrusting others into the same predicament is to restore (somehow) the consolations we have lost.

I turn now to a second Luddite lament, one that views the sciences more generally as dehumanizing. The complaint often takes the form of a critique of technology (or the "technological society") and frequently shows the influence of Marxist ideas, particularly the concept of alienated labor.[2] Sometimes it appears to take a universal form: all of us are diminished by the rise of science and scientism. On other occasions, it seems as though certain privileged people (intellectuals? scientists? bureaucrats?) might still be able to lead worthwhile lives, but that the culture of the masses has been decisively impoverished.

What exactly is the complaint? There's a whole litany of charges. With the rise of the sciences we've ceased to think about questions of values, the efficient attainment of goals has become a goal in itself, we have adopted an alienating world-picture, this world-picture (the one given us by science) is a myth. It's further suggested that the success of natural science has led to the enthronement of scientific reason in the study of human beings, creating the framework for contemporary societies. Finally, the problems that are diagnosed for the sciences are themselves reflections of broader injustices within allegedly democratic societies—science turns out to be the scullion of capitalism.

Not only are there many different charges, but, as is fairly evident, there are several targets. Sometimes the enemy seems to be natural science as a whole. In other cases the trouble comes from a philosophical attitude associated with science (positivism or perhaps scientism) or a decision to apply the findings, concepts, or methods of the natural sciences in a particular human or social domain. Thus, as we'll see, what appears to be a straightforward claim that the

2. For representative versions of the critical stance I have in mind, see Jacques Ellul, *The Technological Society* (New York: Knopf, 1964); Kurt Hübner, *Critique of Scientific Reason* (Chicago: University of Chicago Press, 1983); Max Horkheimer and Theodor Adorno, *Dialectic of Enlightenment* (New York: Continuum, 1994); and Herbert Marcuse, *One-Dimensional Man* (Boston: Beacon, 1991). In various writings, Jürgen Habermas reacts to these discussions in ways that have some points of contact with my own response to them (although I don't claim that I completely understand Habermas' complex views). In thinking about this often opaque literature, I've been greatly helped by Raymond Geuss's lucid and penetrating book, *The Idea of a Critical Theory* (Cambridge: Cambridge University Press, 1981).

advancement of knowledge has been bad for us, turns out to be a tangle of themes, some of which have quite different presuppositions and implications. As we'll also see, some of them are reinforcements of my ideal of well-ordered science.

Start with the suggestion that the rise of modern science has preempted discussion of goals and values. Strictly speaking, the position opposed here is a complex of attitudes towards the natural sciences, rather than the natural sciences themselves. If one believed that the only propositions that could be validated (or refuted) are those that can be assessed by logic, mathematics, and the methods of empirical science, and if one believed further that none of these methods can justify the positive endorsement of a goal or value, then it would follow that there could be no rational support for value judgments. Now the first thesis has sometimes been accepted by thinkers enthusiastic about the sciences (and skeptical of the pretensions of some philosophical traditions), and, since Hume, many people have inclined to the view that, while factual findings might undermine the presuppositions of judgments of value (that is, there can be subversive scientific truths), the inference from facts to values is suspect. So a simple response to the complaint is to declare that positivism is dead, and that we are now liberated from the notion that rational discussion of goals and values is impossible.

Nonetheless, it seems to me that the critics have a deeper point. As I noted in chapter 9, the continuing philosophical focus on the methods that enable an individual to form and justify true beliefs and the concomitant belief in the purity of science support the idea that the most striking developments of human rationality take place in a value-free zone. The failure to articulate a broader normative ideal—like my ideal of well-ordered science—encourages the view that the proper conduct of science can be separated from the mushy considerations (and second-rate rationality) that attend discussions of goals and values. In arguing for the context-dependence of epistemic significance and offering the standard of well-ordered science, I hope to have shown how the rational formulation of goals and values is constitutive of the proper functioning of inquiry, and thus to have reinstated just the kind of discussion the critics have found lacking.

The second charge, that the efficient attainment of goals itself becomes a goal, is a natural extension of the first. Given the positivist conception that rational discussion of goals is impossible, the function of inquiry seems to be to provide information that can be used to promote the widest variety of ends people may happen to adopt. Even though it would be inconsistent to suppose that the efficient pursuit of inquiry is a uniquely rational goal, so that achieving a unified system of scientific laws is paramount, it's easy to see how, given an initial (nonrational) commitment to regard the provision of strategies adaptable to the broadest collection of ends as the optimal state, one might be led to see the efficient development of science and technology as an overriding imperative. In earlier discussions, I've questioned the idea of an ideal, all-purpose system and drawn the moral that judgments about the worthiness of potential projects are integral to the proper practice of the sciences. The critics and I thus agree in recognizing the need for something like the ideal of well-ordered science.

Turn next to the charge that the scientific world-picture is alienating. Here I find a threefold ambiguity. On one interpretation, the criticism may be akin to that expressed by religious opponents of cosmology, evolutionary biology, and kindred sciences: in the wake of these discoveries (or "alleged discoveries" as some would say) we find we can no longer sustain values that were central to our lives. On another reading, the charge may be that the ventures and projects undertaken by the sciences and by technology do not reflect the values of the broad mass of the population, and they fail to do so because there are no channels for the refinement and expression of those values. The last version of the charge introduces an explicitly Marxist element: the practice of inquiry in the societies where it flourishes most visibly serves as a tool of the capitalist system, shaping conditions for the workers that estrange them both from productive activities that would constitute true self-development and from one another.

The themes of the first two readings have already surfaced in previous discussions. The charge that the world-picture offered us by the sciences makes human life inevitably meaningless is no more persuasive from the critical theorist than from the religious fundamentalist, and the response would be analogous to that offered in the first part of this chapter. On the other hand, the second interpretation emphasizes the problem of nonrepresentation that figured in the discussion of well-ordered science (in chapter 10), and here again the critics and I are in agreement. But the third version of the charge offers a more radical point, for it may be seen as denying the legitimacy of accepting the framework of liberal democracy and then trying to articulate an ideal of well-ordered science. If the practice of inquiry in a liberal democracy is contaminated by the features of capitalism, then any attempt to focus on science alone will prove misguided. For the moment, I'll postpone any response to this indictment, which exposes the principal difference between my approach and that of the critics.

Consider next the objection that the scientific world-picture is a myth. On the simplest reading, this is just a blunt assertion of the antirealism whose sources were traced (and, I hope, eradicated) in chapter 2. If my arguments there were cogent, then the central claims of important areas of science are true (or, at least, we have good reason to hold them true), and so, in a straightforward sense, the scientific world-picture is not a myth. There are electrons, but there is no Zeus. Yet, once again, quick dismissal would lose an important point. Chapter 4 endeavored to expose a myth in the vicinity, to wit the idea that the scientific world-picture offers a unique system that describes nature in a privileged way. Another interpretation of the critics construes them as making this more subtle point, emphasizing that the categories in terms of which we describe nature are responses to human interests, and, when inquiry is dominated by the interests of a particular group, there may be alternative classifications, alternative practices, and alternative modifications of the world that might have suited a broader range of human concerns. Once more, this way of taking the charge would lead us back to the normative ideal of well-ordered science, for the myth would be that of supposing there's a uniquely privileged way to describe nature—one independent of the values and projects that human beings happen to have.

The last specific complaint I'll consider alleges that scientific reason is inappropriate for the human sciences. This, too, can mean a number of different things. On the view of "scientific reason" I've favored throughout this essay, the fundamentals of human reasoning are pretty much everywhere the same, whether we're pursuing high theory, tackling mundane practical projects, or thinking about our goals and aims. So the ways in which judgments are justified (and discovered) in the human sciences should be no different from those employed in the natural sciences, apart from the obvious point that certain kinds of experimental manipulations are ethically impermissible in dealing with human beings. Thus, if the critics hold that the human sciences have some distinctive "method" that sets them apart from the natural sciences, I reject the charge, and the disagreement would have to be settled by reviewing candidate methods and seeing if they really are distinct from those employed in other areas of inquiry. But however that turns out, there's no basis for concluding that only the application of some distinctive "method of human science," favored by the critics, will preserve us from social policies that cramp human lives. Alternatively, the criticism may be directed against efforts to incorporate parts of the human sciences within the natural sciences, the kinds of reductionist ventures central to the Unity-of-Science view. On this reading, the critics' target is, once again, a philosophical position about the sciences, and, for reasons given in chapter 6, I'm sympathetic to the criticism. Further, it seems to me eminently possible that reductionist visions might translate into harmful social policies (recall the examples of human sociobiology and kindred endeavors that underlay the discussion of chapter 8).

In reviewing, rather briskly, some of the standard charges made against science and its role in "technological societies," I've attempted to disambiguate provocative claims, rejecting some readings and assimilating others to my own defense of the ideal of well-ordered science. There is, however, a deeper issue, already prefigured in the discussion, which can be posed most sharply by asking if the demand for well-ordered science is sufficiently radical. Earlier I posed the question in an explicitly Marxist format, but it can be raised more generally. Are the failures of the practice of inquiry in democratic societies, the lapses from the state of well-ordered science, really problems about science at all, rather than reflections of broader political problems in those societies? Critics of science, and of my discussion in this book, may propose that the answer to this question is "No," and that my concerns are directed entirely at the wrong level. Restricting one's attention to science, one can pursue the traditional projects about organizing inquiry (exploring the notion of individual rationality, or even collective rationality), but as soon as one lifts one's gaze to the wider values that enter into choices about what lines of inquiry to pursue, there's no stopping short of a full critique of the surrounding society. It's simply a myth to believe that well-ordered science can be promoted without addressing much wider problems.

I have doubts about this line of reasoning at all its stages. First, it's far from obvious that the answer to the question posed in the last paragraph is "No." Precisely because of the gap (chasm?) in knowledge and language that divides the specialized scientist from the outsider, the problems of infusing inquiry with democratic ideals are particularly acute. Indeed, we might expect that those problems might arise even in a society whose commitment to democratic ideals was evident in other facets of its organization, precisely because of the difficulties that motivate the discussions of *Science—The Endless Frontier* and kindred works. Second, even if it's granted that failure to live up to the standard of well-ordered science is partly explained by background social or political inequities, that hardly shows that the task of articulating the appropriate ideal for the practice of inquiry is unimportant. After all, a first step in identifying broad political problems is to pinpoint the various concrete ways in which they manifest themselves, so that recognizing the normative ideal for inquiry might be viewed as a constructive contribution to a broader political critique. Third and last, arguments for the presence of current inequities, as well as plans for removing them, often proceed by starting with a restricted context in which the trouble is particularly clear. Thus a particular institution—education, the media, or even science—can serve as a fulcrum for a wider political project, as we come to appreciate first the need to amend it and then the impossibility of doing so without confronting a more general issue.

The original Luddites smashed machines, and contemporary critics of the sciences sometimes write as if they would like to burn the research laboratories, destroy the equipment, and halt all further inquiry. Since they have no basis for believing that *all* science must make us worse off (for, as I remarked, that is far less plausible than the contrary thesis whose credentials were examined in the last chapter), they must be motivated by the thought that some parts of science are so dreadful that we should not risk developing more of them. I've been suggesting that the awfulness hasn't been demonstrated in any particular instance, and that the sound points in the critical case are incorporated in my plea for well-ordered science, but even if some examples of the terrible consequences of subversive scientific truth withstood scrutiny, the call for an end to research would be an overreaction.

Imagine that we find ourselves in a skeptical mood, seriously concerned that some discoveries have transformed our values in ways that have brought real losses. As we look forward, we want to prevent similar things from occurring again. What exactly should we do? The first thought is that there will probably be some areas of inquiry that frighten us. Perhaps we're anxious about the findings of human behavior genetics or the neuropsychology of the emotions. It's natural then to suppose that our trepidation should express itself in a ban, that we ought to declare certain areas of research off-limits.

If we can foresee that certain types of inquiry might destabilize our scheme of values, then damage has already been done. Losses occur in the transformation of values precisely because we are forced to abandon goals and/or strategies that have given shape and direction to human lives. Now if some project or aspiration is central to a person's life, the idea that the project or aspiration is both valuable and achievable has to carry an especially large motivational burden, so that it can't be a serious question whether presuppositions are satisfied or whether the end is attainable. Yet to identify some areas of research as off-limits is already to express the doubts that would require us to cast around for new ways of making sense of our lives, to start the process of reconstruction that goes on in the wake of subversive discoveries.

Return to one of the examples of the last chapter, the possibility that research might undermine our understanding of friendship, leading us to endorse some vulgar, but true, account of our proclivities and motivation. As things stand, we recognize the bare possibility that some such story is correct, but it doesn't count as a serious possibility for us. Our attitude towards it is of a piece with our recognition that well-entrenched parts of physics, chemistry, or biology might be wrong ("It *might* turn out that covalent bonds don't work through electron-sharing"). If we came to think it possible in a stronger sense, so that it became an open question for us—like the question whether there's intelligent life elsewhere in the universe—then our projects would already be dislocated by doubts about whether the relationships central to them have the characteristics that justify our pursuits. A ban would be pointless, since we'd already be launched on the kinds of value reconstruction that worry us—and it might even be counterproductive, for, after all, the further course of inquiry might *resolve* our doubts.

A second problem for the strategy of trying to protect our schemes of values by banning some types of inquiry is that history shows how disturbing conclusions can come from apparently innocent quarters. Consider Freud's trio of "shocks to our self-esteem": Copernicus' discovery of the earth's motion, Darwin's recognition of our kinship with other animals, and Freud's own claimed disclosure of the workings of the unconscious. How could one have predicted that efforts to reform the Church calendar, to catalogue past and present organisms belonging to exotic faunas and floras, or to address the emotional difficulties of upper-middle-class Viennese women would have such far-reaching consequences? If history is to be used as a guide, however, the most striking moral would be that danger can come from unexpected quarters and that any limited ban would be ineffective.

Perhaps, then, the skeptical meditation should inspire a more radical policy. Perhaps the losses that come from transformations of our schemes of values have been so severe that we'd do better to give up on scientific inquiry entirely. Any such proposal faces the obvious rejoinder that this would be to abandon the real benefits we might gain in terms of advancing our current goals. Further, unless we suppress elaborations and applications of existing knowledge, there's no guarantee that the world has been made safe for the current scheme of values. The skeptical attitude has now become severe, not only supposing that there's no basis for thinking that discoveries that subvert our schemes of values are good for us, but holding that the threat is so dire as to require us henceforth to abandon all thought (there should be nothing new under the sun). Skepticism has turned into conservatism, verging on paranoia, recalling the attitudes of those unfortunate people who refuse to go out because of everyday dangers like crossing the street.

For all the fervor of declarations that the sciences have greatly improved human well-being and the equal ardor with which particular scientific or technological developments have been denounced, we know remarkably little about the effects of inquiry on the quality of lives. Luddite nostalgia for our pastoral past invites salutary reminders about the prevalence of dirt, discomfort, and disease in the alleged Arcadia. Yet there is no systematic evaluation even in the area—medicine—where the positive impact of the sciences would seem most obvious. Medicine can be labeled "the greatest benefit of mankind" but the relationship between the understanding of disease and human well-being, even when construed narrowly in terms of hygiene, health, and longevity, has been quite irregular.[3] Of course, the issues become even more complex when the critics remind us (as they surely will) of the environmental messes bequeathed by a variety of agricultural and technological applications.

It should hardly be surprising that both sides can cite historical examples, and that the record is genuinely mixed. With the best will in the world, people are sometimes unable to foresee the consequences of new knowledge. The complexities of causal linkages typically outrun our comprehension, leading us to

3. See Roy Porter's illuminating survey (which borrows its ironic title from Dr. Johnson), *The Greatest Benefit of Mankind* (New York: Norton, 1997).

intervene in nature in ways that we take to be good for all but that turn out badly.[4] Our predictive ineptness is, however, only part of the trouble. Frequently, the interests of a large portion of our species have been routinely neglected in the agenda-setting and applications of science. Perhaps the champions of the value of knowledge see only benefits because they belong to a group whose lives have genuinely been enhanced, while others, the unrepresented, bear the costs.

Do human lives increase in quality as we come to know more about the natural world, including ourselves? Some say "Yes." Others, pointing to particular discoveries that they regard as damaging, say "No." I say "Maybe—but, above all, it's possible that the variation in the quality of lives at a single historical stage is greater than the variation in quality across time."

The dispute that has dominated the last two chapters is often a clash of rival theologies. The champions of knowledge subscribe to general slogans about the overriding importance of knowing the truth, while their opponents wave banners proclaiming the primacy of other values. I don't think we should subscribe to either of these theologies. We need a different, humbler, picture of the ways in which human welfare changes in the advance of inquiry.

At whatever stage they live, people formulate plans and goals in light of the information they've been able to acquire, and their efforts to carry out the plans and attain the goals depend on the resources available to them. It would be the height of arrogance to believe that only we, the latest characters in the human drama, have managed to identify ends that are genuinely worthwhile, that our freedom in choosing those ends is notably increased, or that our ability to achieve them has been spectacularly enhanced. Instead, I'm inclined to say that there have been many successful experiments in living, many ways in which people have led lives of high quality. Equally, there have always been lives that went badly, people who were enslaved or coerced into particular roles, people who lacked the resources to achieve worthwhile goals that they set for themselves, people blinded by ignorance who pursued trivial or illusory ends. It's important to recognize that the hindrances the unfortunate have encountered are not measured relative to some common early-twenty-first-century state, but to the situations of their luckier contemporaries: the slaves contrast with the free of their own times, those deprived of resources compare with those who were then better endowed, the blinding ignorance is not a matter of failing to appreciate what we know but of lacking knowledge available to others in contemporaneous or preceding societies. If the champions of the value of knowledge are

4. Two excellent sources of examples of our limited ability to recognize the impact of application of knowledge and to guard against disaster are Robert Pool, *Beyond Engineering* (New York: Oxford University Press, 1997), and Edward Tenner, *Why Things Bite Back* (New York: Vintage, 1996).

to propose a plausible thesis, it should be that the advance of inquiry somehow helps to spread the quality of life more broadly, removing the obstacles I've noted. But whether it does so depends crucially on the democratization of science, the extent to which the ideal of well-ordered science is realized.

I ended chapter 11 with the concern that the standard of well-ordered science was too conservative, that it would be at odds with the important idea of the liberating value of truth. In reviewing the debate between those who contend that knowledge is always good for us and those who counter that particular pieces of knowledge are damaging to human well-being, I've arrived at two connected conclusions. The first is that neither side has established its claims, and that we should resist both the grand theology of truth and its antithesis. We need neither the rhetoric of science as the distinctive achievement of human civilization, bracing and liberating in its delivery of truth, nor the gloomy diagnoses of *the* modern predicament. Instead of totalizing "master narratives" about human history and the role of inquiry within it, I commend the humbler picture sketched above, with its commitment to acknowledging human variation.

Second, I've argued that there are no effective strategies for protecting ourselves against the subversion of our values. However we conduct it, the future course of inquiry might force a transformation of the values of a particular group of people that would be difficult and painful. Instead of trying to achieve a level of foresight that is beyond us, we do better to concentrate on the identifiable ways in which inquiry can promote or retard the projects to which specific groups would subscribe after reflective deliberation. Myopic creatures should realize the limits of their vision and adjust their behavior accordingly. What is crucial, of course, is that circumspection should be informed by the interests of the alternative perspectives present in our societies, that the agenda for inquiry and the applications of results should flow from the best approximation we can achieve of the ideal deliberation among representatives of the range of points of view. In the end, then, the winding argument of the last two chapters, with its refusal to honor theology in any of the clashing forms, underscores the importance of the ideal of well-ordered science. That is the best we can hope to do.

FOURTEEN

Research in an Imperfect World

IDEALS ARE ALL VERY WELL, but sometimes the gaps between an ideal and the predicaments in which people find themselves are large enough to make it unclear what should be done. Contemporary scientists don't work in a state of well-ordered science, and the discrepancies are not difficult to appreciate. A recurrent predicament arises because of the opportunity to pursue research that is likely to have consequences of which those who would be most affected are largely unaware, typically bringing benefits of particular kinds to some and injuries of different types to others. Without any process of tutoring the preferences of the vast majority, the research goes forward and scientists compete for the opportunity to play a part in it. Everyone in the community of inquirers knows that the conditions of well-ordered science aren't satisfied, that there are problems of inadequate representation and of false consciousness, but there's a line of defense. It runs like this:

> We can either take this opportunity or pass it up. If we seize it, there will be significant benefits for a large number of people—that is, consequences that those people would welcome if they thought about their needs and wants. It's true that there is a risk of some harm, but we can devise ways of minimizing the harm. Once those protections are in place, even those who would be the victims would concede that the overall program is worth pursuing, that the benefits are too great and the protective efforts, while not failsafe, are adequate. True enough, we haven't made all this clear in advance, but if we had gone through a lengthy explanation, there would have been a public consensus on the advantages of the research we favor, and, in the end, the public good will be enhanced by this program.

This rationale meets an obvious response. Critics of the program argue that the benefits are overblown and the injuries underestimated. Most important, however, they contend that protections to which the rationale gestures, while available in principle, are not likely to be instituted in practice. Because of background social conditions, large features of the society in which the research goes forward and in which its results will be applied, the proposed solutions to the problems are just not realistic. In effect, then, the defense of the program is false advertising, because its ways of counting costs presuppose an absurdly optimistic vision of the conditions of application.

At the next stage of the dialectic, the advocates reply by emphasizing the benefits they expect from the research:

> There are real advantages for a majority of the population—a large majority—and these should not be given up lightly. True enough, there may be accompanying difficulties for a minority. We've pointed out ways in which these difficulties might be overcome. Perhaps our solutions are not politically feasible, but that's hardly the fault of our program. You can't expect scientific inquiry, and the advantages it brings, to be held hostage to the broader problems of society. We should go on, and do the good we can, hoping, of course, that political conditions will change so that the attendant harms may be decreased, rather than letting political considerations dictate the conduct of inquiry.

The last sentence that I've put into the mouth of the program's champions might easily serve as the link to familiar forms of defense. Inquiry is the disinterested search for the fundamental structure of the world; the primary duty of the researcher is to seek the truth, and scientists shouldn't be bound by the temporary conditions of politics; individual investigators are not responsible for the social conditions that render their findings more harmful than they otherwise would have been. The first two of these lines of reaction invoke ideas that previous chapters have attempted to undermine: I've been resolutely opposing any theology of science that would insulate inquiry against moral and political critique. The last builds on a standard image of science to absolve the scientist from blame from the foreseeable harms of research.

What then *are* the responsibilities of scientists, individually and collectively? Given that they live in an imperfect world, far from the conditions of well-ordered science, what ought they to do? In this chapter I'll consider these questions, setting them in the concrete context of the genomes project, an endeavor which, as we've already seen, has a range of social consequences.

The major implications of current genomic research are not hard to predict. Complete knowledge of genome sequences for yeast, worms, flies, and mice will contribute to many projects in theoretical biology, illuminating those large questions about development and physiology that arouse human curiosity. Selective knowledge of the sequences for other organisms, at carefully chosen loci, will clarify issues about evolutionary relationships. The epistemic significance of this venture is evident.

Looking into the distant future, we can reasonably hope that greater understanding of human genetics, human development, and human physiology will enable our descendants to cope better with diseases that currently baffle us. Several decades hence, there may be improved means of treating or preventing cancer, heart disease, diabetes, and other ills of the affluent world. Perhaps there will also be new weapons for combating the infectious diseases that afflict other societies and that may return to haunt Europeans and North Americans as well. Because the road from knowledge of gene sequences to strategies of intervention is blocked at various points by obvious obstacles, it's far from clear to what extent, and for which diseases, the new molecular knowledge will be the key to medical progress. We'd be unlucky if it brought us nothing, but any sober estimate of the future should acknowledge that the benefits are likely to be hard-won and unpredictable in advance.

In the interim, of course, affluent societies will enter the era of genetic testing. People will be able to discover if they are at risk for diseases that might strike later in life; parents will be able to assess the probabilities that newborn infants will develop various conditions; it will be possible to analyze even the fetal genome during pregnancy. Despite the fact that many of the traits with which people are most concerned depend on complicated interactions among genes at many loci and between the genotype and the environment, genetic testing promises probabilistic predictions for a host of traits. Within a decade or so, it will probably be feasible to assess a few thousand loci by drawing a single sample.

In recent years, fledgling biotechnological companies have invested resources in applications of new knowledge about human genetics. The sequencing data can be readily translated into genetic tests, and companies hungry for profits can be expected to market these tests aggressively. Doctors, many of whose understanding of medical genetics is based on vague recollections of a handful of lectures they attended a decade or so ago, will receive brochures touting the benefits of the tests, and, in a climate where allegations of medical malpractice constantly trouble the profession, a significant number will decide to play it safe and recommend that their patients undergo genetic screening. Among their patients, those who are least at home in medical environments, particularly the less privileged, will be swept into taking tests that provide little useful information but that sometimes issue frightening warnings, easily misinterpreted, without any adequate support from genetic counselling. How frequently will such cases arise? Nobody knows. The point, however, is that simply injecting enormously enhanced possibilities of genetic testing into the current socioeconomic context in which medicine is practiced in the affluent world not only allows such scenarios but invites them—we know that biotechnology is a highly competitive business, that genetic tests are easily marketable, that doctors are typically not in a good position to make judgments about the value of such tests, that those doctors face professional risks if they decide not to take "all possible steps," and that a large percentage of patients (perhaps the majority) find the medical environment so alien (and alienating) that they readily submit to following whatever advice their physicians give them.

But there are further troubles. In 1996, at a meeting organized jointly by the NIH/DOE working group on the Ethical, Legal, and Social Implications of the HGP and by the National Action Plan on Breast Cancer, several brave women described their experiences with genetic testing for early onset breast and ovarian cancer.[1] Their stories were grim. Insurers refused to cover prophylactic mastectomies that the women, and their doctors, saw as measures required to avoid grave risks to life. Some women lost their jobs. All had great difficulty in obtaining subsequent health coverage. In order to preserve their anonymity, they spoke under assumed names, and video recording of their presentations was not allowed. Several of the women expressed their distress at being forced to live double lives.

On almost every occasion on which the immediate consequences of genomic research are discussed, some member of the audience either poses a question or comes up to the speaker at the end. The narratives take the same form—a relative or friend took the test for some debilitating (and, most saliently, *expensive*) condition and is now having trouble with insurance or finding a new job. The diseases vary. In the wake of the genomes project we can be sure that there will be many more of them.

The obvious way to forestall the trouble is to insist on "genetic privacy." Insurers and employers cannot be allowed to ask for information about the genotypes of clients and candidates. Yet, as representatives of the insurance industry pointed out at the NIH/DOE/NAPBC meeting, if potential insurees have relevant information that the insurers lack, then the market is in danger of collapse: underwriters try to set competitive premiums (and the spokesmen were very strong on the virtues of competition); in doing so, they will make it attractive for people at high risk to insure themselves heavily, but people who know their risks are low will sign up for less expensive policies; insurers can thus expect that the clients' knowledge will skew the market so that the insurance

1. The tests in question were administered before the sequencing of BRCA1 and BRCA2 and were based on finding familial markers. Since I'll later offer criticisms of the cautious pragmatism which has been pursued by the NIH under Francis Collins's leadership, it's only fair to point out here that he (together with Karen Rothenberg) played a major role in organizing the meeting, and that the women who testified expressed their appreciation of his support, both as a physician and as an advocate of their rights.

companies lose. This economic argument has collided with concerns about the sufferings of those who receive bad news about their genotypes with predictable results. In the United States, there are currently some half-hearted and timorous measures advertised as guaranteeing "genetic privacy," which effectively require insurers to provide some rather unattractive policies to people, irrespective of their genotypes, but which do not solve the problems of those who have genetic predispositions to severe (or expensive?) diseases and disabilities because they involve high premiums or caps on reimbursement (or both). Matters are better with respect to health insurance in other affluent societies, where there are typically government schemes to guarantee some level of medical treatment for all, but questions about life insurance, about higher levels of medical care, and even about access to employment haven't been completely solved.

These are the easy issues. Far more disturbing are the possibilities of using genetic testing prenatally. During the past twenty years, since the development of methods to identify the Tay-Sachs genotype in utero, most parents faced with the diagnosis have chosen to abort, reducing the incidence of this fatal, early onset neurodegenerative disease in the population to less than 1% of the prior rate. It's very hard to see the abortion of Tay-Sachs fetuses as anything other than an act of mercy. But genomic research will multiply, by an enormous factor, the number of cases in which prospective parents will be able to detect the characteristics of embryonic lives. Some of the examples will prove as uncontroversial as Tay-Sachs, for there are other, less well-known, syndromes that afflict young children with severe developmental damage—Lesch-Nyhan syndrome whose bearers are not only profoundly retarded but also have a compulsion to gnaw their lips and fingertips, San Filippo syndrome in which retardation is accompanied with ferocity toward other people, Canavan's disease which involves the same type of neural degeneration found in Tay-Sachs except that it takes about ten years instead of one or two.

Yet there's a host of other possible applications. First come milder types of developmental abnormality with variable phenotypes, as in the case of Down's syndrome, fragile X syndrome, Hurler syndrome, and the like. There are degenerative muscular diseases with genetic bases, as well as forms of blindness, deafness, and dwarfism. Some diseases, like cystic fibrosis, will shorten the child's life-span, involve periods of hospitalization, and rule out many of the normal activities of children and adolescents. Others, like Huntington's and Alzheimer's, will strike late. Genetic testing will be able to identify risks for cancers of many kinds with various times of likely onset, chances of cardiovascular problems, probabilities of contracting diabetes. Beyond these, correlations between genotype and phenotype, *relative to environments that are currently standard,* may enable prospective parents to assess the chances that the fetus would grow into an alcoholic, or a schizophrenic, or a sufferer from bipolar syndrome, or a child with learning difficulties or attentional problems. Perhaps there will be well-established correlations (*again environment-relative*) with diffidence, or antisocial behavior, or low intelligence, or tendencies to outbreaks of violence. Surely genetic testing will enable parents to make educated guesses about body-build, height, eye color, hair type, and so forth. Possibly, there will be the opportunity to assess the probabilities of same-sex preference . . .

So many tests. Which ones should be employed? One tempting response is to declare that the prospect of knowing so much in advance—and acting on the knowledge—is sufficiently daunting that prenatal testing ought to be banned across the board. That response can easily grab at a pejorative name, declaring that the prenatal use of genetic tests is a form of eugenics, and that we should have none of it. Any blanket ban not only overlooks the benign employment of genetic knowledge in sparing families the misery of watching a child decay and die—as with Tay-Sachs and a number of other diseases—but also ignores important distinctions among types of eugenic programs. Eugenics, broadly construed, attempts to modify the characteristics of descendant populations by using genetic knowledge in differential promotion of births. Eugenic practices differ in important ways, most notably in whether they allow freedom of choice to prospective parents or whether they engage in social coercion. When optimists consider prenatal genetic testing, they view it as a benign form of eugenics, one which enables people to forestall outcomes that would bring misery and suffering.

Rosy visions should be scrutinized. Can we seriously believe in these uncoerced decision-makers, able to reflect knowledgeably and compassionately, guaranteed public support for their decisions, whatsoever those should be? If not, then we must come back again to the plethora of genetic tests that will be available and ask what actual people in actual societies are likely to do with the possibility of broad-scale prenatal testing. Within the affluent societies, access to prenatal genetic testing is likely to be quite unequal: perhaps almost all of the population will have the opportunity to take tests for the most severe diseases (although in the United States it's hard to be confident even about that), but the range of possible tests will surely increase, dramatically, with income levels. Many reproductive decisions won't be reflective and uncoerced, because of economic pressures, lack of support, or through the pressure exercised by a prejudiced community. Finally, traditional views about the differential worth of various kinds of people, coupled with the competitive environment of free-market capitalism, make it hard to sustain any broad and tolerant perspective on the ways in which human lives can flourish—indeed contemporary affluent societies are marked by conditions that are likely to channel prenatal genetic testing towards a very narrow ideal of ourselves.

Other regions of the world already make plain how background social prejudices can play a disturbing role in prenatal decisions. In Northern India and in parts of China, clinics that offer amniocentesis, designed to provide couples with the chance to terminate pregnancies when the fetus has chromosomal abnormalities, are regularly visited by women who want to know a different characteristic of the karyotype—not whether it shows an extra (fragment of) chromosome 21 (as in Down's syndrome) but whether it is XX rather than XY. Being female is not a disease, and it ought not to prevent someone from living a life of the highest quality. The women who go to the clinics, seeking to terminate pregnancies with female fetuses, know, however, that in the conditions in which their progeny will grow up, XX is a marker for wretchedness, for neglect, discrimination, maltreatment, and limited horizons. We should not see their reproductive decisions as "free and uncoerced" or suppose that the communities to which they belong permit "broad and tolerant perspectives about human worth."

Those who think that the eugenic consequences of the genomes project in the affluent world will be quite different must believe one of two things, either that the societies in which prenatal genetic testing will become readily available already contain safeguards that will prevent similar deviations, or that advance knowledge of the problems will enable us to design and implement appropriate protections. There's a familiar refrain: "Aren't there ways of avoiding the kinds of outcomes we see in the uses of prenatal testing in India and China, so that we could retain the advantages of the compassionate uses without engaging in unsavory forms of eugenics?" It is reasonable to concede the possibility, in principle, of deploying genetic knowledge to promote a great deal of good while avoiding the obvious evils. Whether what is in principle possible has much to do with what is overwhelmingly probable is quite another matter.

When scientists originally sought governmental funding for genomic research, they were very clear that the results of the project might pose some problems. In an unusual, and admirable, move, James Watson, one of the principal advocates and the person who would become the first director of the NIH genome project, argued that 3% of the funds should be earmarked for investigations of the "ethical, legal, and social implications" of the HGP; the figure was later increased to 5%.[2] So the ELSI program was born.[3]

The hope was that advance knowledge of potential pitfalls would enable policymakers to institute safeguards. It hasn't worked that way. To be sure any number of reports, articles, and books have pinpointed the areas in which increased genetic knowledge can cause trouble. Moreover, it is not hard to devise

2. The DOE project continues to use 3% of its funds for ELSI research.

3. There has recently been some controversy about whether Watson's original intention was to address the potential implications of genomic research or simply to create a buffer that would protect the research from social critics. In *The Clone Age* (New York: Holt, 1999) Lori Andrews cites an unnamed source who claims that Watson wanted the ELSI group to "talk and do nothing"; see chap. 12, especially p. 206.

proposed solutions to many of the difficulties that emerge. Because there are problems about the promotion of unnecessary genetic tests, there should be a regulatory agency to ensure that the exact benefits of newly introduced tests are clearly understood, and that doctors are not misled into thinking that they must require their patients to take tests whose character they (the doctors) don't fully understand. Because more genetic counseling will be needed, and because counseling sessions should be more successful than they are, it's important to invest right now in programs to train people to fill this important role. Because members of some groups are estranged from contemporary practices of medicine, there should be channels to involve those groups, and to help them cope with hospitals and clinics. Because genetic information can be used to set impossibly high premiums or to impose caps on coverage, the United States should follow other affluent nations in making health care available to all—the possibility of genetic discrimination buttresses other arguments for a national health care system. Because similar problems arise with respect to life and disability insurance, effectively depriving the genetically unfortunate of the opportunity to make provisions for their own future and the futures of those whom they love, a parallel program (the "national life security" system?) should be introduced, not only in the United States but also in other affluent societies. Because genetic tests could be used to discriminate in employment, employers ought not to be allowed to request information about candidates or employees (or to treat more favorably people who voluntarily provide it).

I defer, for the moment, the harder problems considered in the last section, problems that stem from our increased powers of prenatal testing. The first point to note is that, even though my formulations are blunt, the lines of solution of the last paragraph have been elaborated at some length in a variety of places.[4] The second point is that the problems are real and that the solutions genuinely address them: women who receive terrible news about their susceptibility to breast cancer wouldn't have to fear the *social* damage to their lives if the proposed systems were in place. Sadly, the third point is that, a decade after the discussions began, virtually nothing has been accomplished—there have been some inadequate efforts to curtail the freedom of insurance providers (without upsetting them too much) and one signal success, an Act of Congress that genuinely reduces the chances of genetic discrimination in employment.[5]

4. Thus Andrews et al., *Assessing Genetic Risks* (Washington D.C.: National Academy Press, 1994), works out the details of ways of assimilating genetic testing. In *The Lives to Come* (New York: Simon & Schuster, 1996), I defend all the proposals I've mentioned.

5. The U.S. Congress achieved this by extending the Americans with Disabilities Act. Ironically, this one success testifies to the principal thesis for which I'll be arguing, to wit the intertwining of problems generated by the genomes project with broader sociopolitical issues. Given the existence of larger commitments to end discrimination in employment, it wasn't hard for politicians to extend them.

There's no great mystery about why things have turned out in this way. If the genomes project had raised isolable problems, then assigning 3–5% of the funds to tackle the social consequences would have been a superbly successful strategy. The trouble is that the principled solutions to the problems affluent societies face in the era of genetic testing require changing prevalent practices in much broader contexts—the problems are entangled with larger sociopolitical questions. Societies that have already decided that they can't afford to invest in their poorest members, that imposing regulations on free enterprise is suspect, that governmental programs typically fail, and so forth are hardly likely to be responsive when they're told that applications of some new science may exacerbate social inequality or make the plight of the poor even worse. Expecting that the prospect of genetic testing will suddenly make the case for universal health care or for some guarantee of security later in life or for easing the access of socially marginalized people to the medical care they need seems politically naïve.[6] To put the point another way, if powerful people were going to be sensitive to principled solutions to the problems raised by genomic research, then they would already have responded to the implications of the pertinent principles in a broad range of contexts, and many of the conditions that make genetic research threatening would already have been removed.

ELSI has failed. Indeed, I think ELSI was doomed to fail from the beginning. Its promise rested on a presupposition that was quite reasonably believed, but that turned out to be false. That presupposition assumed that there was a complex of issues about which politicians of different stripes fight and a bundle of problems raised by genetic investigations that was quite separate—thus one could hope to solve the latter in a politically neutral way. Instead, principled solutions to the problems of genomic research simply became part of broader political debate, coming out on the losing side. All that's left for ELSI is to "take a pragmatic attitude," effectively shuffling the pieces in a very tight space, and perhaps to pretend that something is actually being done to head off the predictable harms widespread genetic testing will bring.[7]

So far I have postponed the harder issues raised by prenatal testing. In a society that tolerates very different levels of access to medical care, we can expect

6. I should plead guilty to at least some measure of political naivete. It did seem to me for a while that considerations of self-interest might move people to rethink the issue of health care coverage. After all, even the descendants of the affluent may turn out to be among the 5–10% of the population who have genotypes that will make it hard or impossible for them to secure health insurance. On purely prudential grounds, a national health care system is an excellent bet. But, within the United States, negative attitudes toward governmental programs are so entrenched that I suspect that many people will have to suffer before the obvious solution is finally accepted.

7. This is perhaps an appropriate place to note the hard work and insight of many people who have contributed to the ELSI program—I think of Nancy Wexler, Lori Andrews, Troy Duster, Dan Drell, Tony Holtzmann, David Cox, Dorothy Nelkin, and Rebecca Eisenberg. They should no more be blamed for ELSI's lack of success than the defenders of more famous lost causes.

that more affluent parents will have the option to take a wider range of prenatal tests than the less well off. Quite possibly this will lead to a situation in which various kinds of diseases and disabilities will be far more prevalent among the poor, with the result that public support for programs to enhance the lives of those who have such diseases and disabilities withers. Furthermore, residues of traditional prejudices are likely to affect the practice of prenatal testing. During the past few decades, men who are sexually attracted to other men, and women sexually attracted to other women, have made some progress in some places with respect to some aspects of their lives. It's abundantly clear, however, that, in most affluent societies, homosexuals are vulnerable to violence and humiliation and that they have to struggle to obtain the benefits routinely allowed to heterosexuals. The kind of sex selection practiced in Northern India and in China is not likely to be widespread in affluent societies (although chromosomal tests have been used by Western women desperate for a boy—or a girl). Nevertheless, prejudice against homosexuals, and the more muted forms of prejudice against fat people and short people, is likely to fuel prenatal testing for "gay genes" (or "genes for obesity," or "short" genes). Even those who have no prejudice against homosexuals may react to their perception of prejudice around them: like the Northern Indian women who recognize the brutalities visited on girls, they perceive that the attitudes of members of their communities diminishes the quality of life of homosexuals, and, in consequence, they regretfully take steps to ensure that their children have the best chance, that sons don't have a higher probability of being gay, daughters an increased probability of being lesbian.

In many affluent societies, most notably in Britain and the United States, but increasingly in societies that used to have a firm commitment to taking care of the disadvantaged, there is a large gap between those who have the most resources and those who have least. The steeper the pyramid, the greater the pressure on parents committed to the principle of doing what they can for their children to invest in a wide variety of prenatal tests. In a society that offers meaningful work at reasonable wages to people whose performances on I.Q. scores is about a standard deviation below the mean, so that such people, on average, live about as well as those who test at a standard deviation above the mean, there would be no great interest in "genes for I.Q." (more exactly, genotypes that correlate in prevailing environments with scores on particular tests[8]).

8. It's important not to accept conjectures about the *causal* roles genes play in the development of intelligence or about the *invariance* of phenotypes across a broad range of environments. But, given the zeal with which researchers look for "genetic bases" for intelligence, it's quite likely that they will discover correlations between some genotypes and high (or low) levels of test performance in children who are reared in standard suburban home and school environments. That research can then provide information to the effect that one fertilized egg has a 30% greater chance than another of developing into a child whose test score falls into some desired range, provided that the child is

Elsewhere, in places where the differences in life prospects between people of these two types are enormous, parents who have been successful in the competition will want their children to inherit the advantages, and that goal can be promoted by selecting fetuses who pass the prenatal test.

Earlier, I argued that the issues raised by the genomes project could not be neatly separated from broader concerns. That is even more evident with respect to prenatal testing. Tackling the issues in this arena would require taking steps to eradicate residual prejudices, making a commitment to creating a society in which people with many different traits could live well, establishing programs of support for those who need special environments in which to develop, equalizing access to medical care, and using the resources of molecular behavioral genetics to avoid leaping to seductive conclusions about genetic determinants and seriously trying to map the environmental factors that contribute. If there are identifiable ways of solving the relatively easy problems considered earlier (instituting universal health care coverage, a program for life and disability insurance, and so forth), the changes that are now required are far more global, and, consequently, far less likely to occur.

Sad to say, there's little chance that the ELSI program will even take the first steps with respect to these more difficult questions. From the beginning of the project, NIH and DOE instituted a "joint working group" that would consider the issues raised, discuss projects that would be worth funding, and "set an ELSI agenda." This group contained committed and knowledgeable people from a number of different fields.[9] In 1997, the working group proposed investigating the implications of work in molecular behavioral genetics and asked for funds to commission a series of studies. At this point, NIH announced that no such funds were available, that the working group's activities had to be curtailed for budgetary reasons, and, after a period of discussion, the chair of the working group resigned.[10] In the wake of her resignation, it was decided that there should be a committee review of the working group, and, after the review took place, the working group was disbanded, with suggestions that its responsibilities would be taken over within a new structure.[11]

---

reared in the kind of environment typical of the socioeconomic class of the parents. Even without lapsing into genetic determinism, behavior genetics can thus provide probabilistic information prenatally.

9. The first chair, Nancy Wexler, is well known for her genetic research and, in particular, for her fieldwork that culminated in the mapping of the Huntington's locus. The second, Lori Andrews, is a lawyer who has contributed greatly to our understanding of the uses of genetic tests for late onset diseases. The third, Troy Duster, is an eminent sociologist who has explored the eugenic implications of genomic research.

10. The chair was Lori Andrews. For her account of the episode, see *The Clone Age*, chap. 12.

11. I had been a member of the working group for about a year at the time of Andrews's resignation, and I know that opinion among the members was divided, some viewing the NIH as trying to squelch concerns that weren't welcome, others reserving judgment. Although the group was

Modifying the structure of a joint enterprise is often a good thing—as people see how the work is evolving they can improve on the arrangements they made in advance. The judgment should be made by considering whether the changes promote the goals of the venture, in this instance whether the investigations of ELSI issues are improved and whether the results of those investigations are more readily translated into policy. But it seems that the ELSI program is currently at a standstill and that there have been no notable successes in coping with the imminent problems. To put the point a bit uncharitably, the net result of the disbanding of the working group seems to have been to prevent a handful of serious scholars from opening up issues leaders of the genomes project would rather not have aired.

It's no secret that NIH prefers pragmatism to principle. The mandate for the ELSI program seems to be to find ways of accommodating genetic research to prevailing political constraints, doing the best that the mood of Congress allows rather than trying to articulate the fundamental problems and convey them clearly to politicians and public alike. Pragmatism may be a sensible strategy, a means of doing the little good one can instead of campaigning for a broader set of changes that might undermine a project that will prove its value in the long run. Yet, quite clearly, pragmatism can become the counsel of the comfortable, a way of avoiding confronting genuine harms and injustices, and can even decline to the state in which ELSI is simply a showcase, in place just to defuse criticism of the genomes project.[12] When harmful consequences ensue (or threaten), champions of genetic research can point to the ELSI program as a sign of their social concern—"We do the best we can to solve these difficult problems."

What, then, are the responsibilities of contemporary molecular biologists, particularly those whose research is currently funded by the genomes project? Imagine a thoughtful genome scientist who has followed carefully the discussions of the impact of the genetic research from the beginning. Our scientist agrees, give or take a quibble or two, with the analysis I've offered. As Jean (to give her a name) looks into the future, she foresees important medical advances at the end of the twenty-first century, achieved as a result of the work in which she and her colleagues are now engaged, but she also predicts that, unless there

---

a joint NIH/DOE group, there was no doubt that the principal tension was between the group and NIH.

12. In defense of NIH, one might note that, at the end of the 1990s, the publicly funded project was threatened by the possibility that an almost-complete human genome sequence would be generated by a private company (Craig Venter's *Celera*). The scientific challenge surely made ELSI seem even more peripheral to the NIH mission, and it would have been reasonable to claim that retaining control of the project was necessary for *any* attention to ethical, legal, and social implications. The private project comes with no ethical strings attached.

are serious changes in the sociopolitical climate, the actual implementation of new genetic knowledge in the next twenty years will do more harm than good to the most vulnerable members of her society. She considers a number of possible courses of action. One would simply be to wind up her current genomic research and go back to work on the immune system of mice. But she knows that, even if she abandons the human research, the work will still be done in other laboratories. Perhaps, then, she ought to play a more active role, persuading her colleagues to join her in abandoning the project? Again, she recognizes that the mapping and sequencing of the human genome will continue even if her talented friends jump ship. So perhaps she ought to take a more radical stance, arguing publicly for the cessation of genomic research until the sociopolitical conditions exist in which research can be applied without the foreseeable harms? This line of action also disturbs her, in part because she doubts her ability to succeed in doing much more than angering colleagues whose judgment she values, in part because she thinks of scientific inquiry as delicate and vulnerable and fears that, were she to be successful, the likely result would be a disastrous overreaction (withdrawal of public support for many kinds of benign projects, for example). Because she dislikes the alternatives, she decides to continue with the human research, scrupulously avoiding entanglements with biotechnology companies, attending workshops on ELSI issues, and speaking out in favor of measures she views as limiting the potential harms, and even writing the occasional article about the need to address the problems of the eugenic future. From time to time, she wonders whether her quietist stance is too timid and, occasionally, when she reads about Vichy France (or listens to Tom Lehrer), her twinges of conscience become sharp pangs.

Jean's predicament isn't unprecedented. Many people find themselves occupying roles in institutions whose overall contributions they value, even though there are aspects of those institutions they regard as fostering injustice or other forms of harm. Radical action appears quixotic, since its effect will be to deprive the person of the opportunity to pursue the valuable goals associated with the role, and the role will be filled by someone else so that the bad consequences will occur in any case. More moderate ways of proceeding leave the discomfiting thought that one has simply acquiesced in wrongdoing. So, for example, Jean's position isn't very different from that of people who work for firms committed to exploiting some of their employees (perhaps in distant lands) or that of officers of institutions that don't take steps to remedy past social injustices (as in the case of universities that close their affirmative action programs).

Let's identify some salient features of the situation. First, an agent currently fills a position such that the actions required by that position can be expected to produce negative consequences for a class of people currently disadvantaged or marginalized within society. Second, any alternative action available to the agent has negligible probability of avoiding those consequences (or, perhaps, any alternative with nonnegligible probability of avoiding the consequences would cause far more harm). Third, alternative forms of action would be likely to decrease the probability that the larger venture in which the position is embedded would realize its valuable goals. It will be useful to give names to these three aspects of the predicament: the first is a condition of *predictable harm*, the second is a condition of *impotence* with respect to that harm, and the third is a condition of *institutional entanglement*.

The obvious source of impotence is the willingness of others to fill the position if the agent refuses to go along with its demands, but that willingness may itself be fueled by the institutional entanglement—people who recognize that biomedical research, taken as a whole, contributes to human welfare may be able to come to terms with the prospect of carrying out the inquiries with predictable harm. The source of the institutional entanglement is, I suggest, traceable to two lapses, first the gap between the actual practice of inquiry and well-ordered science and second a broader failure of democratic commitment.

To see why this is so, let's suppose we were in a state of well-ordered science. Then the decisions to pursue lines of research and to apply them in ways that would bring the predictable harms would reflect the consensus of ideal deliberators formed after the tutoring of preferences. That could only occur if, even in the wake of the expression of the needs of the historically disadvantaged, a majority of the population believed that such needs could legitimately be overridden. It's possible, of course, that basic preferences of the majority are insensitive to the injustices suffered by others, so that, even after tutoring, the majority would continue to insist on projects that bring predictable harm to the disadvantaged. In this case, however, we have an even more basic departure from democratic ideals, in that there's a fundamental refusal to take the plight of some other members of society seriously, and that refusal is, I suggest, incompatible with a commitment to regarding those individuals as fellow-members of a democratic society. Hence, if the transition to well-ordered science failed to remedy the predicament then that failure would result from a deeper lapse from the norms of democracy.

We naturally think that the obligations of scientists must center directly on their practice of the particular lines of research in which they're engaged, that it's a matter of going ahead or abstaining from investigations according to the ways in which they're likely to promote or retard human well-being, especially for those who are most vulnerable. In light of the suggestions of the last paragraphs, I want to offer a broader conception of responsibilities. Scientists have the obligation to do what they can to nudge the practice of inquiry in their society closer to the state of well-ordered science, and citizens have the responsibility to do what they can better to approximate democratic ideals.[13] The enterprises in

13. This discussion is indebted to Amy Gutmann's views about our responsibilities to promote racial justice. See *Color Conscious*, co-authored with Anthony Appiah (Princeton: Princeton University Press, 1996), 172–174.

which people engage typically have both narrow and broad functions—thus the narrow function of an educational institution is to teach the students, but it also has the broader function of contributing to a just society. In the case that concerns us, the narrow function of the sciences is to generate significant truths, where the criteria of significance are those in force within the community of inquirers. The broad function is to promote the democratic practice of science, as conceived in the ideal of well-ordered science (so that, for example, the criteria of significance actually in force represent the outcome of an ideal deliberation among ideal agents). From this broad function flows the responsibility to attempt to lessen the gap between actual practice and the ideal. Beyond that, at an even wider level, the scientist, as citizen, also has the obligation to do whatever is possible to realize more fully democratic ideals in her society.

Back to Jean. Her discomfort stems from the fact that her research is implicated in predictable harm under conditions of impotence and institutional entanglement. She can do nothing constructive directly. Yet the recognition of her predicament and of its causes should, if I'm right, lead her to make efforts to discharge her broader responsibilities as scientist and as citizen. That is, she can take steps to address the basic problems that generate situations like hers, by trying to move her community towards a state in which its practice will be closer to that of well-ordered science. I'll now try to explain, speculatively and tentatively, what kinds of specific things might be done to meet these broader responsibilities.

The first maxim for a responsible scientist is a classic, commending self-knowledge. It's all too easy, I think, for those deeply engaged in a research project to overlook any harms to which that project might contribute. To decrease the gap between the current practice of inquiry and the state of well-ordered science, one should start by recognizing the existence and character of the gap. Thus, for someone now working on the genomes project, it's important not to pretend that the public rationale for the project identifies the real reasons for undertaking it, important also to appreciate the fact that the large benefits are likely to be distant and the genuine ills more immediate, as well as to see that the task of avoiding those ills isn't something that can confidently be left to others. My imaginary scientist, Jean, has already come this far.

Suppose, then, that Jean has a clear view of the ways in which genomic research lapses from the state of well-ordered science. What, concretely, can she do to address those lapses? Start with the issue of false advertisement. The simplest thought is that Jean has the obligation to proclaim the truth, to point out that the promise of cures to be developed quickly is an illusion, and to explain the indirect reasons why genomic research is worth pursuing. Although this option isn't subject to the condition of impotence—Jean doesn't need others to join her to launch a successful exposé—there are still difficulties of institutional entanglement, and Jean may reasonably fear that only the negative part of her message will be heard. But if Jean believes that the ground needs to be prepared for an open discussion that will avoid the tyranny of the ignorant,[14] then she has the obligation to do some of the digging. In particular, she should use what skills she has to advance public understanding of the questions with which she's concerned and to encourage people outside science to appreciate the point of the inquiries she and her colleagues undertake. One aspect of this responsibility is supporting those in her community who attempt to articulate scientific ideas in popular settings (provided, of course, that they do so accurately); another is to think creatively about forms of education in science that would aim to give students a broad understanding of how particular fields hang together rather than serving as the early phases of an initiation into research science.[15] Furthermore, insofar as the benefits of the projected investigations are epistemic, they deserve to be appreciated more broadly, and the task of improving public understanding of the sciences can be expected to stimulate and satisfy the curiosity of many people who are not involved with the fine details.

When there's an identifiable group of people who will be adversely affected by the applications of the research in which the scientist is engaged, then there's a special responsibility to do whatever can be done to acquaint members of this group with the potential harms and to give voice to their concerns. One of the great ironies of genomic research is that families in which particular diseases are inherited have often been enthusiastic supporters of the inquiries, contributing time and information that help the scientists who take an interest in "their" disease to identify the locus, even though, as the families surely do not know, the likely result of the labors will be a genetic test that may prove burdensome for people very like themselves.[16] Besides the obligation to inform affected groups about the likely consequences of genomic research, scientists also have the responsibility to help such people make their concerns vivid for a broader public, perhaps by giving their time to show that the anxieties are quite genuine, perhaps by persuading elected representatives who champion the interests of the pertinent groups.

14. It's quite clear that some biologists believe that there's a stronger obligation to level with the public now, come what may. This is surely the perspective of Lewontin's "The Dream of the Human Genome," in his *Biology as Ideology* (New York: Harper, 1992). In the text, I'm concerned to identify a minimal set of responsibilities rather than trying to decide just how far the obligations go.

15. Of course, specialized courses from high school through postgraduate education will always be necessary, but it seems to me a terrible mistake not to recognize that, from quite early stages in their academic careers, some students need not "rigorous introductions" to physics, biology, chemistry, etc., but courses designed to introduce the main ideas and to allow outsiders to engage in the kinds of discussions that well-ordered science envisages. Citizens needs to know the kinds of things about the sciences that will enable the advocates of research programs to explain their rationales clearly and openly.

16. I should note that some researchers have gone to considerable lengths to support the families who have helped them. Nancy Wexler's extraordinary tirelessness in promoting the health (and other interests) of Venezuelan Indians at risk for Huntington's is exemplary.

But perhaps the most obvious thing that Jean can do to move her community closer to a state of well-ordered science is to take the step that initially seemed so quixotic, to abandon, very publicly, her funding, and channel her efforts towards other forms of research. Especially if she is well-known, and especially if she is respected by her colleagues, a public renunciation of the project on the clearly stated grounds that too little has been done to protect the underprivileged against foreseeable harms might spur the public discussion and the policy-setting that have been so conspicuously lacking. The rationale for quitting would be somewhat different from that implicit in my earlier discussion—Jean would understand that the research would still be done—but she would aim to prepare for a closer approximation of well-ordered science.

I offer these suggestions tentatively and speculatively, recognizing that different investigators will find themselves in quite different situations, with varied opportunities for promoting the state of well-ordered science. The fundamental point I've been making is that researchers have obligations, individually and collectively, to work in ways that approximate the requirements of well-ordered science, and, where this is impossible, because of impotence and institutional entanglement, to do what they can to bring the practice of science closer to well-ordered science. Those obligations are especially clear at a time when scientific research is increasingly co-opted by entrepreneurs whose interest in profits is likely to have little to do with the tutored collective preferences of other citizens. But to speculate further would be to venture into the details of politics and to enter domains where I must confess my ignorance. I have been arguing for the scientist's responsibility to engage in sociopolitical reflection and to let that reflection inform one's actions, but exactly how to give content to the directive is a matter for policy-makers to pursue in the large and for individual scientists, aware of the nuances of their situations, to decide. Here, then, I cease.

Afterword

By way of conclusion, it may be helpful to summarize the argument I have offered in this essay. Our contemporary discussions of the sciences are divided between two false images, one, that of the faithful, which views inquiry as liberating, practically beneficial, and the greatest achievement of human civilization, the other, that of the detractors, which sees science as an expression of power, a secular religion with no claims to "truth," which systematically excludes the voices and the interests of the greater part of the species. I've been urging something different, a view that derives points from both camps, sometimes modifying them and fitting them together into what I hope is a coherent whole.

I began by defending the notion of truth, and justifying the idea that the sciences sometimes deliver the truth, even about quite recondite entities and properties. A modest realism, I claimed, survives even the most sophisticated challenges. Yet there's an important insight in the constructivist opposition. The notion of significance, I argued, is not independent of time and context, and our standards of epistemic significance, like our standards of practical significance, evolve, not aiming toward any predetermined end, but being modified by the practices and institutions of the past. Indeed, there's a clear sense in which the sciences are constitutive of our world, the mundane sense in which what we pick out as important and worthy of investigation at one time leads to interactions with the environment that modify it for our successors. Nature is shaped by our past interests, its current configurations partially determine our present needs, and out of these needs grow our further attempts to solve problems we take to be epistemically and practically significant.

The upshot of this is that epistemic values do not stand apart from—

"above"—our quotidian concerns. Rather, they are to be balanced against practical interests. Attempts to pose certain issues and to answer them can interfere with human well-being, or with the welfare of those who have been historically disadvantaged. In consequence, the abstract ideals that are often brought to discussions of the sciences, based as they are on a neat separation between the epistemic and the practical, are sometimes inappropriate; I offered as an example the hallowed notion of free inquiry. The very type of liberalism that has traditionally motivated the ideal of free inquiry reveals its limitations when certain practical conditions fail.

Once this point has been appreciated, the next step is to consider what an appropriate ideal for the practice of science might be. I endeavored to show how the traditional conception of the function of science as aiming at truth should be expanded, and I proposed the notion of well-ordered science. In the ideal of well-ordered science, truth retains a place, but it is set within a democratic framework that takes the proper notion of scientific significance to be that which would emerge from ideal deliberation among ideal agents.

Because well-ordered science might seem too conservative, not appreciating the power of inquiry to liberate us from our prejudices, I developed and defended my ideal by considering the processes through which values are transformed. I suggested that there is no general reason for believing that processes of scientific enlightenment inevitably conduce to human well-being, nor are there grounds for thinking that the transitions that have actually occurred in the past have somehow diminished human lives. My aim was to debunk both the theology of science, with its professions that science is a high calling dedicated to ends that transcend all others, and to oppose the counter theology, the demonization of science. Once again, after rejecting theology, we find ourselves with a conception in which the epistemic and the practical, truth and democracy, intertwine.

The final step in the argument was to consider the problems posed by lapses from the ideal and the responsibilities of those who work on projects that conflict with the ideal. Using the genomes project as an example, I offered an analysis of why it is proving so hard to integrate this part of science with reflective social aspirations. This analysis then served as the basis for a sketch of how responsible scientists might behave, identifying an obligation to do whatever can be done to bring about the condition of well-ordered science. This transition between the construction of a philosophical framework—my principal task—and the crafting of practical policies and decisions is, I admit, radically incomplete, but I hope that others, with greater competence in the pertinent fields of social science, will see how to go further.

The position to which I've been led is not a comfortable one. It's likely to inspire the charge, from detractors, that my introduction of social and political issues is too timid and conservative, and the accusation, from boosters, that science is too precious and too important to flirt with democracy. To the former, I can only offer the invitation to reflect on my arguments, to consider the possibility that the metaphysics and epistemologies conjured up in opposition to the claims of the sciences are simply botched attempts to formulate the kinds of ethical, social, and political concerns I've articulated, and to analyze the totalizing rhetoric of the sweeping condemnations of science. To the latter, it's worth pointing out that it would be a curious irony to insist that the sciences represent the epitome of human reason, while simultaneously defending their insulation from public discussion, however prefaced with attempts at enlightenment, because the voices of unreason will inevitably triumph. If for no other reason, democracy in science, exemplified in well-ordered science, is worth aiming for because, like democracy generally, it is preferable to the alternatives.

Essay on Sources

Introduction

For the "Science Wars" of the 1990s, see Paul Gross and Norman Levitt, *Higher Superstition* (Baltimore, Md.: Johns Hopkins University Press, 1994); Andrew Ross, ed., *Science Wars* (Durham, N.C.: Duke University Press, 1996); and Noretta Koertge, ed., *A House Built on Sand* (New York: Oxford University Press, 1998). Enthusiastic defenses of science, in the tradition of T. H. Huxley and John Tyndall, can be found in Carl Sagan, *The Demon-Haunted World* (New York: Ballantine, 1996), and E. O. Wilson, *Consilience* (Cambridge, Mass.: Harvard University Press, 1998). In *Whose Science? Whose Knowledge?* (Ithaca, N.Y.: Cornell University Press, 1991), Sandra Harding suggests that science as practiced is confining and that we need inquiries specifically designed to meet the needs of women, children, minorities in affluent societies, and people from developing countries. A more general treatment of science as a center of power and oppression is offered by Michel Foucault in *Power/Knowledge* (New York: Pantheon, 1980).

Chapter One

The positions of the faithful and the debunkers are represented in the works cited in connection with the *Introduction*. Here, and throughout the book, my view of the Human Genome Project is a descendant of the perspective I articulated in *The Lives to Come* (New York: Simon & Schuster, 1996); I have also benefited from Robert Pollack, *The Missing Moment* (Boston: Houghton Mifflin, 1999) (see especially the appendix), and Lee Silver, *Remaking Eden* (London: Weidenfeld and Nicholson, 1998). Steven Weinberg's *Dreams of a Final Theory* (New York: Vintage, 1994) is useful for appreciating the appeal of the superconducting supercollider. In considering biological arguments about behavioral differences I have learned much from Ned Block and Gerald Dworkin, eds., *The IQ Debate* (London: Quartet, 1977), from Block's essay "How Heritability Misleads about Race," *Cognition*, 56, 1995, 99–128, and from Richard Lewontin, Steven Rose, and Leon Kamin, *Not in Our Genes* (New York: Pantheon, 1984); I have also drawn on my study of human sociobiology in *Vaulting Ambition* (Cambridge, Mass.: MIT Press, 1985). A recent account of resistance to Darwinism among contemporary Creationists is Robert Pennock, *Tower of Babel* (Cambridge, Mass.: MIT Press, 1999).

Chapter Two

This chapter takes up issues that have been much discussed by recent historians and sociologists of science, as well as by professional philosophers. Many of the arguments opposed in early parts of the chapter are advanced by such writers as Steven Shapin, *A Social History of Truth* (Chicago: University of Chicago Press, 1996); David Bloor, *Knowledge and Social Imagery* (London: Sage, 1974); Harry Collins, *Changing Order* (London: Sage, 1985); and Bruno Latour, *Science in Action* (Cambridge, Mass.: Harvard University Press, 1987); and *We Have Never Been Modern* (Cambridge, Mass.: Harvard University Press, 1993). The original form of the argument from the inaccessibility of reality stems from early modern writers such as Locke, *Essay Concerning Human Understanding* (reprint New York: Dover, 1963); my response to this argument summarizes points well developed by J. L. Austin in *Sense and Sensibilia* (Oxford: Oxford University Press, 1962); by George Pitcher in *A Theory of Perception* (Princeton: Princeton University Press, 1971); and by Jonathan Bennett in *Locke, Berkeley, Hume: Central Themes* (Oxford: Oxford University Press, 1971).

The appeal to the history of science to undermine realist views is begun in Thomas Kuhn, *The Structure of Scientific Revolutions* (Chicago: University of Chicago Press, 1962/1970), but the argument that successful prediction doesn't betoken the truth (or approximate truth) of the theory that yields the predictive successes is most fully developed by Larry Laudan in "A Confutation of Convergent Realism" (most readily available as the final chapter of his *Science and Values* [Berkeley: University of California Press, 1984]). In chapter 5 of *The Advancement of Science* (New York: Oxford University Press, 1993), I offered a reply to this argument, which is elaborated here. Empiricist arguments that methods of justification must be subject to checking are advanced by Bas van Fraassen in *The Scientific Image* (Oxford: Oxford University Press, 1980) and by Arthur Fine in *The Shaky Game* (Chicago: University of Chicago Press, 1986). My discussion of Galileo is much indebted to Albert van Helden, *The Invention of the Telescope* (Chicago: University of Chicago Press, 1985), and *Galileo's Sidereus Nuncius*, edited by van Helden (Chicago: University of Chicago Press, 1989). Arguments similar to my defense of realism have been offered by Richard Miller, in *Fact and Explanation* (Princeton: Princeton University Press, 1987), and by Jerrold Aronson, in "Testing for Convergent Realism," *British Journal for the Philosophy of Science*, 40, 1989, 255–259.

The challenge that realism is committed to unacceptable metaphysics is presented forcefully by Arthur Fine, *The Shaky Game*; Hilary Putnam, *Reason, Truth, and History* (Cambridge: Cambridge University Press, 1981); Nelson Goodman, *Ways of Worldmaking* (Indianapolis: Hackett, 1978); and Richard Rorty, *Consequences of Pragmatism* (Minnesota: University of Minnesota Press, 1981). The discussion of the text outlines my response to these writers, while omitting some details that would lead into side issues without, I think, affecting the general position. I offer more detailed presentations in two forthcoming essays: "Real Realism: The Galilean Strategy" (*Philosophical Review*, 2001, in press) and "On The Explanatory Power of Correspondence Truth" (to appear in *Philosophy and Phenomenological Research*).

Chapter Three

Challenges to objectivity that proceed via the underdetermination of theory by evidence typically draw on the final section of W. V. Quine, "Two Dogmas of Empiricism," in *From a Logical Point of View* (New York: Harper, 1956). For influential recent versions, see Steven Shapin and Simon Schaffer, *Leviathan and the Air-Pump*; Harry Collins, *Changing Order*; and Helen Longino, *Science as Social Knowledge* (Princeton: Princeton University Press, 1990). The examples from physics are illuminatingly discussed in Bas van Fraassen, *The Scientific Image*; in Larry Sklar, *Space, Time, and Space-Time* (Berkeley: University of California Press, 1974) and "Saving the Noumena," in his *Philosophy and Spacetime Physics* (Berkeley: University of California Press, 1985). Recent attempts to respond to underdetermination arguments include Larry Laudan, "Demystifying Underdetermination," in *Scientific Theories*, ed. C. Wade Savage (Minneapolis: University of Minnesota Press, 1990, 267–297), and Laudan and Jarrett Leplin, "Empirical Equivalence and Underdetermination," *Journal of Philosophy*, 88, 1991, 449–472. A lucid overview of the many ways in which the notions of empirical equivalence and underdetermination can be conceived is John Earman, "Underdetermination, Realism, and Reason," *Midwest Studies in Philosophy*, 18, 1993, 19–38. My response to Kuhnian concerns about the impotence of evidence in revolutionary scientific debates is elaborated at much greater length in chapter 7 of *The Advancement of Science*.

Chapter Four

My concerns about the difficulties of finding a useful account of "natural kinds" have been inspired partly by the scruples of W. V. Quine, "Natural Kinds," in his *Ontological Relativity* (New York: Columbia University Press, 1970), and Nelson Goodman, *Fact, Fiction, and Forecast* (Indianapolis: Bobbs-Merrill, 1956); partly by John Dupré, *The Disorder of Things* (Cambridge, Mass.: Harvard University Press, 1993), and Ian Hacking, *The Social Construction . . . of What?* (Cambridge, Mass.: Harvard University Press, 1999). In insisting that the modest realism of chapters 2 and 3 isn't committed to the idea of some privileged language for describing reality, I try to distinguish theses that Richard Rorty, *Consequences of Pragmatism,* and Hilary Putnam, *The Many Faces of Realism* (La Salle: Open Court, 1987), tend to view as an inextricable whole. The position of this chapter has several important points of contact with Nelson Goodman's *Ways of Worldmaking,* although, set against the background of chapter 2, it lacks the implication that Goodman's critics have found repugnant (see, for example, Israel Scheffler, "The Wonderful Worlds of Goodman," *Synthese,* 45, 1980, 201–209). The idea that many different languages might deliver the same truths, while "parsing" the world differently is the central theme of W. V. Quine, "Ontological Relativity" (in *Ontological Relativity*), although my version of this is an "explosive" form of realism in the sense of Ernest Sosa, "Putnam's Pragmatic Realism," *Journal of Philosophy,* 90, 1993, 605–626.

For the many-sided debate about species concepts, see Marc Ereshevsky, ed., *The Units of Evolution* (Cambridge, Mass.: MIT Press, 1992). My pluralist views are defended in "Species" and in "Some Puzzles about Species"; an important brief for pluralism that has influenced the discussion of the chapter is John Dupré, "Natural Kinds and Biological Taxa," *Philosophical Review,* 90, 1981, 66–90. In thinking about natural kind terms in the physical sciences, I have greatly benefited from discussions with Kyle Stanford; our shared conclusions are advanced in "Refining the Causal Theory of Reference for Natural Kind Terms," *Philosophical Studies,* 97, 2000, 99–129, which elaborates further the perspective I offer here. The thought that our classifications might affect the world in the mundane way outlined at the end of the chapter has been clarified by some of Peter Godfrey-Smith's presently unpublished work on John Dewey's "realism."

Chapter Five

My account of maps and their features has been informed by David Greenhood, *Mapping* (Chicago: University of Chicago Press, 1964); Mark Monmonier, *How to Lie with Maps* (Chicago: University of Chicago Press, 1991); and Arthur H. Robinson, *Early Thematic Mapping in the History of Cartography* (Chicago: University of Chicago Press, 1982). In every sense, my eyes have been opened by three books by Edward Tufte: *The Visual Display of Quantitative Information* (Cheshire, Conn.: Graphics Press, 1983), *Envisioning Information* (Cheshire, Conn.: Graphics Press, 1990), and *Visual Explanation* (Cheshire, Conn.: Graphics Press, 1997). A more elaborate account of my view of maps is offered in a joint article with Achille Varzi, "Some Pictures Are Worth $2^{\aleph_0}$ Sentences," *Philosophy*, 75, 2000, 377–381. Besides Stephen Toulmin's comparison of maps to scientific theories in *Philosophy of Science: An Introduction* (New York: Harper and Row, 1953), I am also in sympathy with the analogy suggested by John Ziman in *Real Science* (Cambridge: Cambridge University Press, 2000) (see section 6.5).

Chapter Six

On confusion about the significance of cloning, see Richard Lewontin, "The Confusion Over Cloning" and my "Life after Dolly," both in Glenn McGee, ed., *The Human Cloning Debate* (Berkeley: Berkeley Hills Books, 1998). The *locus classicus* of the idea that many truths are completely insignificant is Karl Popper, *The Logic of Scientific Discovery* (London: Hutchinson, 1959). For representative versions of various views about the epistemic aims of inquiry, see Ernest Nagel, *The Structure of Science* (New York: Harcourt Brace, 1961); C. G. Hempel, *Aspects of Scientific Explanation* (New York: Free Press, 1965); Wesley Salmon, *Scientific Explanation and the Causal Structure of the World* (Princeton: Princeton University Press, 1984). Bas van Fraassen opposes these conceptions of the aims of the sciences in *The Scientific Image* and *Laws and Necessity* (Oxford: Oxford University Press, 1989). My position shares some features with van Fraassen's critique, and perhaps more with the views elaborated by Nancy Cartwright (in *The Dappled World*) and John Dupré (*The Disorder of Things*). My differences with Cartwright and Dupré are explored in "Unification as a Regulative Ideal," *Philosophical Perspectives*, 7, 1999, 337–348. The overall position of the chapter is also much closer to that offered by Larry Laudan in *Science and Values* than was the perspective taken in *The Advancement of Science*; but though I agree with Laudan that our scientific goals evolve, I see this more in terms of concrete projects than the kinds of large epistemic desiderata on which he focuses.

The idea of an "explanatory store" was bruited in my "Explanatory Unification," *Philosophy of Science*, 48, 1981, 507–531, in connection with an account of explanation I now see as at least partly misguided, but I think the notion of the explanatory store has an obvious wider usage. The classic conception of the notion of reduction that informs the Unity-of-Science View is offered by Ernest Nagel (*The Structure of Science*, chapter 12). An influential attack on the view is Jerry Fodor "Special Sciences," *Synthese*, 28, 1974, 77–115; I've also criticized the view at some length in "1953 and All That. A Tale of Two Sciences," *Philosophi-*

cal Review, 93, 1984, 335–373, and "The Hegemony of Molecular Biology," *Biology and Philosophy*, 14, 1999, 195–210, essays on which the present chapter draws. In "Genes Made Molecular," *Philosophy of Science*, 61, 1994, 163–185, C. Kenneth Waters attempts to provide an account of genes that will serve the reductionist's turn, but, besides limiting the variety of genes, Waters's proposal only offers a functional specification, not the kind of structural description that will mesh with the principles of chemistry.

The idea of nature as a patchwork of laws is defended by Nancy Cartwright (*The Dappled World*), and the pragmatic character of the notion of a scientific law has recently been emphasized by Sandra Mitchell, "Dimensions of Scientific Law," *Philosophy of Science*, 67, 2000, 242–265. Sylvain Bromberger's "Why-Questions," in *Mind and Cosmos*, ed. R. Colodny (Pittsburgh: University of Pittsburgh Press), noted the ways in which philosophers of science have often limited the kinds of explanations they consider. The discussion of relevance relations, and the connection to pragmatic questions about explanation, have been influenced by Bas van Fraassen, *The Scientific Image*; as noted in the chapter, it now seems to me that the joint critique of van Fraassen by Wesley Salmon and me ("Van Fraassen on Explanation," *Journal of Philosophy*, 84, 1987, 315–330) overlooked an important possibility. The idea that ideal explanations trace the complete causal history is suggested by Peter Railton in "Probability, Explanation, and Information," *Synthese*, 48, 1981, 233–256. The notion of a significance graph is an attempt to refine an approach to scientific significance I offered in chapter 4 of *The Advancement of Science*.

Chapter Seven

The discussions of this chapter have been much influenced by Ilkka Niiniluoto, "The Aim and Structure of Applied Science," *Erkenntnis*, 38, 1993, 1–21; Donald Stokes, *Pasteur's Quadrant* (Washington, D.C.: Brookings Institute Press, 1997); and Edward Shils, ed., *Criteria for Scientific Development: Public Policy and National Goals* (Cambridge, Mass.: MIT Press, 1968) (see especially the essays by Michael Polyani, Alvin Weinberg, and Stephen Toulmin). Many authors have recently denied that there's an important distinction between science and technology (see, for example, Bruno Latour's discussion of "technoscience" in *Science in Action*). I think my emphasis on the idea that the category of "pure science" is supposed to defuse certain moral objections provides me with a somewhat different perspective on these issues.

Chapter Eight

A more detailed (and formal) version of the argument of this chapter is provided in my article "An Argument about Free Inquiry," *Noûs*, 31, 1997, 279–306.

My reading of Mill has been influenced by Alan Ryan's "Mill in a Liberal Landscape," in J. Skorupski, ed., *The Cambridge Companion to Mill* (Cambridge: Cambridge University Press, 1998), and Isaiah Berlin's "John Stuart Mill and the Ends of Life," in *Four Essays on Liberty* (Oxford: Oxford University Press, 1969).

The argument that human sociobiology ought to meet high standards of evidence is advanced by Stephen Jay Gould in "Biological Potentiality vs. Biological Determinism," in *Ever Since Darwin* (New York: Norton, 1977), and also in my *Vaulting Ambition* (Cambridge, Mass.: MIT Press), 8–10. The formulation of the political asymmetry has been motivated by some of the suggestions offered by Arthur Jensen, "How Much Can We Boost I.Q.?," reprinted in *The IQ Debate*, ed. N. J. Block and Gerald Dworkin, and by Richard Herrnstein and Charles Murray, *The Bell Curve* (New York: Free Press, 1994). In thinking about the epistemic asymmetry, I've drawn on historical analyses offered by Londa Schiebinger in *The Mind Has No Sex?* (Cambridge, Mass.: Harvard University Press, 1989); Stephen Jay Gould in *The Mismeasure of Man* (New York: Norton, 1981); and several essays collected in Block and Dworkin, eds., *The IQ Debate*; also influential have been methodological diagnoses by Richard Lewontin, "The Analysis of Variance and the Analysis of Causes," reprinted in *The IQ Debate*; Lewontin, Steven Rose, and Leon Kamin, *Not in Our Genes* (New York: Pantheon, 1984), and Ned Block, "How I.Q. Misleads about Race," *Cognition*, 56, 1995, 99–128. In addition, I've generalized points from my critique of human sociobiology in *Vaulting Ambition* and in "Developmental Decomposition and the Future of Human Behavioral Ecology," *Philosophy of Science*, 57, 1990, 96–117.

In formulating the moral point that the crucial issue turns on the expected utility *for the disadvantaged*, I try to absorb well-known criticisms of consequentialism (see, for example, Bernard Williams's contribution to J. J. C. Smart and Williams, *Utilitarianism: For and Against* (Cambridge: Cambridge University Press, 1973); my suggestion is obviously in line with Rawls's famous "difference principle" (*A Theory of Justice* [Cambridge, Mass.: Harvard University Press, 1971]). J. Philippe Rushton's descriptions of his motivations for continuing his research into racial differences serve as a paradigm of the attempt to assume the role of the persecuted scientist: see his *Race, Evolution, and Behavior: A Life History Perspective* (New Brunswick: Transaction Publishers, 1995). My views about duties to seek truth have been aided by comments from Susan Dwyer, and the discussion of the idea that we should promote human reflective deliberation owes much to conversations with David Brink; here I have been influenced by Joshua Cohen, "Freedom of Expression," *Philosophy and Public Affairs*, 22, 1993, 207–263, and Brink's "Millian Principles, Freedom of Expression, and Hate Speech," *Legal Theory*, 7, 2001, 119–57.

In using E. O. Wilson's appeal to liberal values as a point of departure in this chapter, I want to draw attention to a point I should have made explicitly in

Vaulting Ambition: humane and liberal impulses are certainly evident in Wilson's writings (in contrast with those of others who advance similar conclusions) and his work on human social behavior shows, I believe, how even well-meaning researchers of great integrity may offer flawed defenses of socially harmful conjectures.

Chapter Nine

An influential attack on the idea of a method of discovery is chapter 2 of C. G. Hempel, *Philosophy of Natural Science* (Englewood Cliffs, N.J.: Prentice-Hall, 1966). For attempts to show that methods of discovery are possible, see Clark Glymour et al., *Discovering Causal Structure* (Orlando: Academic Press, 1987), and Kevin Kelly, *The Logic of Reliable Inquiry* (New York: Oxford University Press, 1996). Discussions of tradeoffs between gaining valuable information and risking error have been discussed by a number of philosophers, but perhaps most thoroughly by Isaac Levi in *Gambling with Truth* (New York: Knopf, 1967), *The Enterprise of Knowledge* (Cambridge, Mass.: MIT Press, 1983), and *The Fixation of Belief and Its Undoing* (Cambridge: Cambridge University Press, 1991); my formulations of the epistemological issues in this chapter have been aided by discussions with Levi.

The suggestion that homogeneity is bad epistemic policy was defended at some length by Paul Feyerabend; see especially "Against Method," in *Analyses of Theories and Methods of Physics and Psychology*, ed. M. Radner and S. Winokur (Minneapolis: University of Minnesota Press, 1970), 17–130, in which the connection with Mill is explicitly made. I have tried to provide precise analyses of the value of cognitive diversity and the ways in which it might be achieved, in "The Division of Cognitive Labor," *Journal of Philosophy*, 87, 1990, 5–22, and in chapter 8 of *The Advancement of Science*. My rustic efforts have now been refined by William Brock and Stephen Durlauf in "A Formal Model of Theory Choice in Science," *Economic Theory*, 14, 1999, 113–130. For worries about searches for optima in evolutionary inquiry, see Stephen Jay Gould and Richard Lewontin, "The Spandrels of San Marco and the Panglossian Paradigm: A Critique of the Adaptationist Programme," *Proceedings of the Royal Society*, B 205, 1979, 581–598. The general view of social epistemology as an attempt to consider the consequences of social institutions and to try to improve their epistemic effects is advanced by Alvin Goldman in *Knowledge in a Social World* (New York: Oxford University Press, 1999).

A resourceful defense of objectivism about values is given by Thomas Hurka in *Perfectionism* (New York: Oxford University Press, 1993), which contains some useful insights about the possibility of balancing some values against others. In "Essence and Perfection," *Ethics*, 110, 1999, 59–83, I delineate my reservations about Hurka's objectivist project. Part IV of Derek Parfit's *Reasons and*

Persons (Oxford: Oxford University Press, 1984) exposes the difficulty of proceeding from an account of the individual good to an account of the collective good. I try to show how severe the difficulty is in "Parfit's Puzzle," *Noûs*, 34, 2000, 550–577. In considering a subjectivist alternative, I've been greatly helped by appendix I to Parfit's *Reasons and Persons*; James Griffin's *Well-Being* (Oxford: Oxford University Press, 1986); and Peter Railton's "Facts and Values," *Philosophical Topics*, 14, 1986, 5–31.

Chapter Ten

The general conception of well-ordered science owes an obvious debt to John Rawls, *A Theory of Justice* (Cambridge, Mass.: Harvard University Press, 1971) and "Outline of a Decision Procedure for Ethics," reprinted in *Collected Papers of John Rawls*, ed. S. Freeman (Cambridge, Mass.: Harvard University Press, 1999). I've also been influenced by Robert Dahl, *A Preface to Democratic Theory* (Chicago: University of Chicago Press, 1963), and by Amy Gutmann and Dennis Thompson, *Deliberative Democracy* (Cambridge, Mass.: Harvard University Press, 1996). The notion of "tutored preferences" is akin to the idea of refined desires introduced in attempts to distinguish between desires and interests (see, for example, James Griffin, *Well-Being*). In considering the possibility that the sciences might constitute an arena governed by "public morality," I have drawn from Stuart Hampshire, *Innocence and Experience* (Cambridge, Mass.: Harvard University Press, 1989). For illuminating discussion of the problems attending social choice, see Amartya Sen, *Choice, Welfare, and Measurement* (Cambridge, Mass.: Harvard University Press, 1982). The possibility that the decisions made even under ideal circumstances may be parochial was forcefully presented to me by Peter Singer, and analogous arguments appear in Singer's "Famine, Affluence, and Morality," *Philosophy and Public Affairs*, 1, 1982, 229–243, as well as in Peter Unger, *Living High and Letting Die* (New York: Oxford University Press, 1996).

The problem of inadequate representation has been well posed by Helen Longino in her *Science as Social Knowledge*; indeed, I view the chapter both as developing Longino's call for a more democratic organization of inquiry and as raising the question of whether the absence of some groups from some discussions signals a divergence between the actual conduct of inquiry and its ideal pursuit. For a cogent presentation of the Problem of Parochial Application (although they do not employ this concept), see Ruth Hubbard and Elijah Wald, *Exploding the Gene Myth* (Boston: Beacon, 1993). I begin a mathematical analysis of the possible advantages that might accrue from enlightened democracy in "Social Psychology and the Theory of Science" (to appear in Peter Carruthers and Stephen Stich, eds., *The Cognitive Science of Science*, Cambridge University Press).

Chapter Eleven

Besides the three main documents considered in this chapter—Francis Bacon, *New Atlantis* (Oxford: Oxford University Press [World's Classics], 1966); Vannevar Bush, *Science—The Endless Frontier* (Washington, D.C.: NSF, 40th anniversary ed., 1990); and Leon Rosenberg, *Scientific Opportunities and Public Needs* (Washington, D.C.: NIH, 1998)—I have also benefited from a number of other sources. Daniel Kevles's accounts of the Bush report, both in his preface to the reprinting of *Science: The Endless Frontier* and in *The Physicists* (Cambridge, Mass.: Harvard University Press, 1971), have been extremely helpful. (I am also grateful to Kevles for supplying me with a copy of the reprint of the Bush report.) Richard Nelson has offered expert guidance on the recent history of science policy in the United States, and I owe to him the suggestion that I should look at the 1960s debate about goals. My approach, especially in latter parts of the chapter, has been informed by all the contributions to Edward Shils, ed., *Criteria for Scientific Development: Public Policy and National Goals* (Cambridge, Mass.: MIT Press, 1968), by Harvey Brooks, *The Government of Science* (Cambridge, Mass.: MIT Press, 1968), and by Don K. Price, *The Scientific Estate* (Cambridge, Mass.: Harvard University Press, 1967).

Chapter Twelve

After completing an earlier version of this chapter, I discovered Isaiah Berlin's essay "'From Hope and Fear Set Free,'" *The Proper Study of Mankind* (New York: Farrar Straus Giroux, 1998), which contains many similarities of theme and has greatly influenced the final version. One of the few other philosophers to have seriously raised the question of the value of scientific knowledge is Paul Feyerabend; see the closing section of "Against Method," and *Farewell to Reason* (London: Verso, 1987). Although the approach I've adopted shares some similarities with that taken by Feyerabend, a comparison of the arguments of chapters 8 and 13 with his discussions in the works cited and in *Science in a Free Society* (London: New Left Books, 1978) will reveal important differences.

For the crises of Victorian faith, see Leonard Huxley, ed., *Life and Letters of Thomas Henry Huxley* (London: Macmillan, 1900); Adrian Desmond, *Huxley* (Reading, Mass.: Addison-Wesley, 1997); Frank Turner, *Between Science and Religion* (New Haven: Yale University Press, 1974); Bernard Lightman, ed., *Victorian Science in Context* (Chicago: University of Chicago Press, 1997); and Richard J. Helmstadter and Bernard Lightman, eds., *Victorian Faith in Crisis* (Stanford: Stanford University Press, 1990). In *Consilience*, E. O. Wilson offers a very clear development of the dismissive response to complaints about subversive knowledge; Richard Dawkins's *Unweaving the Rainbow* (Boston: Houghton Mifflin, 1998) is slightly more sympathetic to such complaints, but he too assumes that, properly understood, the natural sciences can supply everything the complainants are looking for. My discussion of plans, central desires, and the quality of life is much influenced by John Rawls, *A Theory of Justice* (especially section 63); Derek Parfit, *Reasons and Persons* (part IV and appendix I); James Griffin, *Well-Being*; and Peter Railton, "Facts and Values." A lucid discussion of the balancing of distinct goods is offered by Thomas Hurka in *Perfectionism.* Hurka also tries to provide a grounding for an Aristotelian version of objectivism. In "Essence and Perfection," which further articulates the argument of the closing pages of the chapter, I try to show that the attempt fails.

Chapter Thirteen

There are many classic and modern sources of what I describe as the "Luddites' Laments." See, for example, William Blake, *Jerusalem*; Carolyn Merchant, *The Death of Nature* (New York: Harper, 1983); Max Horkheimer and Theodor Adorno, *The Dialectic of Enlightenment* (New York: Continuum, 1994); Sandra Harding, *Whose Science? Whose Knowledge?* (Ithaca, N.Y.: Cornell University Press, 1991); Herbert Marcuse, *One-Dimensional Man* (Boston: Beacon, 1991); Jacques Ellul, *The Technological Society* (New York: Knopf, 1964); Martin Heidegger, *The Question Concerning Technology* (New York: Harper and Row, 1977); and Kurt Hübner, *Critique of Technological Reason* (Chicago: University of Chicago Press, 1983). As noted in the chapter, my understanding of the work of the Frankfurt school has been much enhanced by Raymond Geuss's *The Idea of a Critical Theory* (Cambridge: Cambridge University Press, 1981).

Chapter Fourteen

In writing this chapter I have further elaborated (with some changes) the views expressed in my earlier writings about the genomes project, particularly in *The Lives to Come* (New York: Simon & Schuster, 1996) and in "Utopian Eugenics and Social Inequality," in P. Sloan, ed., *Controlling Our Destinies* (Notre Dame: University of Notre Dame Press, 2000), 229–262. For different perspectives, which have also influenced me, see Richard Lewontin, *Biology as Ideology* (New York: Harper, 1992); Ruth Hubbard and Elijah Wald, *Exploding the Gene Myth* (Boston: Beacon, 1993); Lee Silver, *Remaking Eden* (London: Weidenfeld and Nicolson, 1998); and Allen Buchanan et al., *From Chance to Choice* (Cambridge: Cambridge University Press, 2000). Lori Andrews's *The Clone Age* (New York: Holt, 1999) offers her perspective on the work of the ELSI program.

In her contribution to *Color Conscious,* co-authored with Anthony Appiah (Princeton: Princeton University Press, 1996), Amy Gutmann makes points about the responsibilities of professionals to promote broader values which I adapt to the present example.

Index